Advances in Intelligent Systems and Computing

Volume 1003

Series Editor

Janusz Kacprzyk, Systems Research Institute, Polish Academy of Sciences, Warsaw, Poland

Advisory Editors

Nikhil R. Pal, Indian Statistical Institute, Kolkata, India

Rafael Bello Perez, Faculty of Mathematics, Physics and Computing, Universidad Central de Las Villas, Santa Clara, Cuba

Emilio S. Corchado, University of Salamanca, Salamanca, Spain

Hani Hagras, School of Computer Science & Electronic Engineering, University of Essex, Colchester, UK

László T. Kóczy, Department of Automation, Széchenyi István University, Gyor, Hungary

Vladik Kreinovich, Department of Computer Science, University of Texas at El Paso, El Paso, TX, USA

Chin-Teng Lin, Department of Electrical Engineering, National Chiao Tung University, Hsinchu, Taiwan

Jie Lu, Faculty of Engineering and Information Technology, University of Technology Sydney, Sydney, NSW, Australia

Patricia Melin, Graduate Program of Computer Science, Tijuana Institute of Technology, Tijuana, Mexico

Nadia Nedjah, Department of Electronics Engineering, University of Rio de Janeiro, Rio de Janeiro, Brazil

Ngoc Thanh Nguyen, Faculty of Computer Science and Management, Wrocław University of Technology, Wrocław, Poland

Jun Wang, Department of Mechanical and Automation Engineering, The Chinese University of Hong Kong, Shatin, Hong Kong

The series "Advances in Intelligent Systems and Computing" contains publications on theory, applications, and design methods of Intelligent Systems and Intelligent Computing. Virtually all disciplines such as engineering, natural sciences, computer and information science, ICT, economics, business, e-commerce, environment, healthcare, life science are covered. The list of topics spans all the areas of modern intelligent systems and computing such as: computational intelligence, soft computing including neural networks, fuzzy systems, evolutionary computing and the fusion of these paradigms, social intelligence, ambient intelligence, computational neuroscience, artificial life, virtual worlds and society, cognitive science and systems, Perception and Vision, DNA and immune based systems, self-organizing and adaptive systems, e-Learning and teaching, human-centered and human-centric computing, recommender systems, intelligent control, robotics and mechatronics including human-machine teaming, knowledge-based paradigms, learning paradigms, machine ethics, intelligent data analysis, knowledge management, intelligent agents, intelligent decision making and support, intelligent network security, trust management, interactive entertainment, Web intelligence and multimedia.

The publications within "Advances in Intelligent Systems and Computing" are primarily proceedings of important conferences, symposia and congresses. They cover significant recent developments in the field, both of a foundational and applicable character. An important characteristic feature of the series is the short publication time and world-wide distribution. This permits a rapid and broad dissemination of research results.

** Indexing: The books of this series are submitted to ISI Proceedings, EI-Compendex, DBLP, SCOPUS, Google Scholar and Springerlink **

More information about this series at http://www.springer.com/series/11156

Francisco Herrera · Kenji Matsui ·
Sara Rodríguez-González

Editors

Distributed Computing and Artificial Intelligence, 16th International Conference

 Springer

Editors
Francisco Herrera🆔
Department of Computer Science
and Artificial Intelligence
University of Granada, ETS de Ingenierias
Informática y de Telecomunicación
Granada, Spain

Kenji Matsui
Department of Engineering
Osaka Institute of Technology
Osaka, Japan

Sara Rodríguez-González🆔
Department of Computer Science
University of Salamanca
Salamanca, Spain

ISSN 2194-5357   ISSN 2194-5365 (electronic)
Advances in Intelligent Systems and Computing
ISBN 978-3-030-23886-5   ISBN 978-3-030-23887-2 (eBook)
https://doi.org/10.1007/978-3-030-23887-2

This Springer imprint is published by the registered company Springer Nature Switzerland AG
The registered company address is: Gewerbestrasse 11, 6330 Cham, Switzerland

Preface

Research on intelligent distributed systems has matured during the last decade, and many effective applications are now deployed. The artificial intelligence is changing our society. Its application in distributed environments, such as the Internet of Things (IoT), electronic commerce, mobile communications, wireless devices, distributed computing is increasing and is becoming an element of high added value and economic potential, both industrial and research. Nowadays, these technologies are changing constantly as a result of the large research and technical effort being undertaken in both universities and businesses. Most computing systems from personal laptops/computers to cluster/grid/cloud computing systems are available for parallel and distributed computing. Distributed computing performs an increasingly important role in modern signal/data processing, information fusion, and electronics engineering (e.g., electronic commerce, mobile communications, and wireless devices). Particularly, applying artificial intelligence in distributed environments is becoming an element of high added value and economic potential.

The 16th International Symposium on Distributed Computing and Artificial Intelligence 2019 (DCAI 2019) is a forum to present applications of innovative techniques for solving complex problems in these areas. The exchange of ideas between scientists and technicians from both academic and business areas is essential to facilitate the development of systems that meet the demands of today's society. The technology transfer in this field is still a challenge and for that reason this type of contributions will be specially considered in this symposium. This conference is the forum in which to present application of innovative techniques to complex problems. This year's technical program will present both high quality and diversity, with contributions in well-established and evolving areas of research. Specifically, 52 papers were submitted by the authors from 21 different countries (Algeria, Angola, Argentina, Brazil, China, Colombia, Denmark, Ecuador, France, Hong Kong, Iran, Japan, Mexico, Netherlands, Poland, Portugal, Spain, Turkey, UK, USA, and Uzbekistan), representing a truly "wide area network" of research activity. The DCAI'19 technical program has selected 29 papers and, as in past editions, it will be special issues in JCR-ranked journals such as Information Fusion, Neurocomputing, Sensors, Processes, and Electronics. These special issues will cover extended versions of the most highly regarded works. Moreover, DCAI'19 special sessions have been a very useful tool in order to complement the regular program with new or emerging topics of particular interest to the participating community.

This symposium is organized by the Osaka Institute of Technology, Hiroshima University, University of Granada, and University of Salamanca and will be held in Ávila, Spain, from June 26 to 28, 2019. We thank the sponsors: IEEE Systems Man and Cybernetics Society Spain Section Chapter and the IEEE Spain Section (Technical Co-Sponsor), IBM, Indra, Viewnext, Global Exchange, AEPIA, APPIA and AIR institute. The funding to support the Junta de Castilla y León (Spain) with the project *"Virtual-Ledgers-Tecnologías DLT/Blockchain y Cripto-IOT sobre organizaciones virtuales de agentes ligeros y su aplicación en la eficiencia en el transporte de última milla"* (ID. SA267P18—project co-financed with FEDER funds); and finally the Local Organization members and the Program Committee members for their hard work, which was essential for the success of DCAI'19.

June 2019   Francisco Herrera
                                                                                Kenji Matsui
                                                                             Sara Rodríguez

Organization

Honorary Chairman

Masataka Inoue   President of Osaka Institute of Technology, Japan
Sigeru Omatu   Hiroshima University, Japan

Program Committee Chairs

Francisco Herrera   University of Granada, Spain
Kenji Matsui   Osaka Institute of Technology, Japan
Sara Rodríguez   University of Salamanca, Spain

Workshop Chair

Enrique Herrera Viedma   University of Granada, Spain

Scientific Committee

Reza Abrishambaf   Miami University, USA
Ana Almeida   ISEP-IPP, Portugal
Gustavo Almeida   Instituto Federal do Espírito Santo, Brazil
Giner Alor Hernandez   Instituto Tecnologico de Orizaba, México
Cesar Analide   University of Minho, Portugal
Luis Antunes   GUESS/LabMAg/Univ. Lisboa, Portugal
Fidel Aznar   Universidad de Alicante, Spain
Zbigniew Banaszak   Warsaw University of Technology,
                                  Faculty of Management, Dept. of Business
                                  Informatics, Germany
Olfa Belkahla Driss   University of Manouba, Tunisia

Alberto Fernandez	CETINIA. University Rey Juan Carlos, Spain
Peter Forbrig	University of Rostock, Germany
Felix Freitag	Universitat Politècnica de Catalunya, Spain
Toru Fujinaka	Hiroshima University, Japan
Bogdan Gabrys	University of Technology Sydney, Australia
Svitlana Galeshchuk	Nova Southeastern University, USA
Jesús García	University Carlos III Madrid, Spain
Francisco Garcia-Sanchez	University of Murcia, Spain
Marisol García-Valls	Universidad Carlos III de Madrid, Spain
Irina Georgescu	Academy of Economic Studies, Romania
Abdallah Ghourabi	Higher School of Telecommunications SupCom, Tunisia
Joseph Giampapa	Carnegie Mellon University, USA
Ana Belén Gil González	University of Salamanca, Spain
Arkadiusz Gola	Lublin University of Technology, Poland
Juan Gomez Romero	University of Granada, Spain
Carina Gonzalez	Universidad de La Laguna, Spain
Evelio Gonzalez	Universidad de La Laguna, Spain
Angélica González Arrieta	Universidad de Salamanca, Spain
David Griol	Universidad Carlos III de Madrid, Spain
Felipe Hernández Perlines	Universidad de Castilla-La Mancha, Spain
Aurélie Hurault	IRIT - ENSEEIHT, France
Elisa Huzita	State University of Maringa, Brazil
Gustavo Isaza	University of Caldas, Colombia
Patricia Jiménez	Universidad de Huelva, Spain
Bo Noerregaard Joergensen	University of Southern Denmark, Denmark
Vicente Julian	Universitat Politècnica de València, Spain
Geylani Kardas	Ege University International Computer Institute, Turkey
Amin Khan	UiT The Arctic University of Norway, Tromsø, Norway
Naoufel Khayati	COSMOS Laboratory - ENSI, Tunisia
Egons Lavendelis	Riga Technical University, Latvian
Rosalia Laza	Universidad de Vigo, Spain
Tiancheng Li	Northwestern Polytechnical University, China
Johan Lilius	Abo Akademi University, Finland
Faraón Llorens-Largo	Universidad de Alicante, Spain
Ivan Lopez-Arevalo	Cinvestav - Tamaulipas, México
Daniel López-Sánchez	BISITE, Spain
Ramdane Maamri	LIRE laboratory UC Constantine2-Abdelhamid Mehri, Algeria
Benedita Malheiro	Instituto Superior de Engenharia do Porto, Portugal
Eleni Mangina	UCD, Ireland
Fabio Marques	University of Aveiro, Portugal

Goreti Marreiros	ISEP/IPP-GECAD, Portugal
Angel Martin Del Rey	Department of Applied Mathematics, Universidad de Salamanca, Spain
Jesus Martin-Vaquero	University of Salamanca, Spain
Fabio Martinelli	IIT-CNR, Italy
Ester Martinez-Martin	Universidad de Alicante, Spain
Philippe Mathieu	University of Lille 1, France
Kenji Matsui	Osaka Institute of Technology, Japan
Shimpei Matsumoto	Hiroshima Institute of Technology, Japan
Jacopo Mauro	University of Oslo, Norway
Rene Meier	Lucerne University of Applied Sciences, Switzerland
José Méndez	University of Vigo, Spain
Heman Mohabeer	Charles Telfair Institute, Australia
Mohd Saberi Mohamad	Universiti Malaysia Kelantan, Malaysia
Jose M. Molina	Universidad Carlos III de Madrid, Spain
Miguel Molina-Solana	Data Science Institute - Imperial College London, UK
Stefania Monica	Università degli Studi di Parma, Italy
Naoki Mori	Osaka Prefecture University, Japan
Paulo Moura Oliveira	UTAD University, Portugal
Paulo Mourao	University of Minho, Portugal
Muhammad Marwan Muhammad Fuad	Coventry University, UK
Susana Muñoz Hernández	Universidad Politécnica de Madrid, Spain
Antonio J. R. Neves	University of Aveiro, Portugal
Jose Neves	University of Minho, Portugal
Julio Cesar Nievola	Pontifícia Universidade Católica do Paraná - PUCPR Programa de Pós Graduação em Informática Aplicada, Brazil
Nadia Nouali-Taboudjemat	CERIST, Algeria
Paulo Novais	University of Minho, Portugal
José Luis Oliveira	University of Aveiro, Portugal
Tiago Oliveira	National Institute of Informatics, Japan
Sigeru Omatu	Osaka Institute of Technology, Japan
Mauricio Orozco-Alzate	Universidad Nacional de Colombia, Colombia
Sascha Ossowski	University Rey Juan Carlos, Spain
Miguel Angel Patricio	Universidad Carlos III de Madrid, Spain
Juan Pavón	Universidad Complutense de Madrid, Spain
Reyes Pavón	University of Vigo, Spain
Pawel Pawlewski	Poznan University of Technology, Poland
Cristina Pelayo	University of Oviedo, Spain
Diego Hernán Peluffo-Ordoñez	Yachay Tech, Ecuador
Stefan-Gheorghe Pentiuc	University Stefan cel Mare Suceava, Romania

Antonio Pereira	Escola Superior de Tecnologia e Gestão do IPLeiria, Portugal
Tiago Pinto	University of Salamanca, Spain
Julio Ponce	Universidad Autónoma de Aguascalientes, México
Juan-Luis Posadas-Yague	Universitat Politècnica de València, Spain
Jose-Luis Poza-Luján	Universitat Politècnica de València, Spain
Isabel Praça	GECAD/ISEP, Portugal
Radu-Emil Precup	Politehnica University of Timisoara, Romania
Francisco A. Pujol	Specialized Processor Architectures Lab, DTIC, EPS, University of Alicante, Spain
Mar Pujol	Universidad de Alicante, Spain
Araceli Queiruga-Dios	Department of Applied Mathematics, Universidad de Salamanca, Spain
Mariano Raboso Mateos	Facultad de Informática - Universidad Pontificia de Salamanca, Spain
Miguel Rebollo	Universitat Politècnica de València, Spain
Manuel Resinas	University of Seville, Spain
Jaime A. Rincon	Universitat Politècnica de València, Spain
Ramon Rizo	Universidad de Alicante, Spain
Sergi Robles	Universitat Autònoma de Barcelona, Spain
Sara Rodríguez	University of Salamanca, Spain
Cristian Aaron Rodriguez Enriquez	Instituto Tecnológico de Orizaba, México
Luiz Romao	Univille, México
Abdulraqeb S. A. Alhammadi	Faculty of Engineering, Multimedia University, Cyberjaya, Selangor, Malaysia
Gustavo Santos-Garcia	Universidad de Salamanca, Spain
Ichiro Satoh	National Institute of Informatics, Japan
Yann Secq	Université Lille I, France
Ali Selamat	Universiti Teknologi Malaysia, Malaysia
Emilio Serrano	Universidad Politécnica de Madrid, Spain
Mina Sheikhalishahi	Consiglio Nazionale delle Ricerche, Italy
Amin Shokri Gazafroudi	Universidad de Salamanca, Spain
Fábio Silva	University of Minho, Portugal
Nuno Silva	DEI & GECAD - ISEP - IPP, Portugal
Paweł Sitek	Kielce University of Technology, Poland
Pedro Sousa	University of Minho, Portugal
Masaru Teranishi	Hiroshima Institute of Technology, Japan
Adrià Torrens Urrutia	Universitat Rovira i Virgili, Spain
Leandro Tortosa	University of Alicante, Spain
Volodymyr Turchenko	Research Institute for Intelligent Computing Systems, Ternopil National Economic University, Ucrania
Miki Ueno	Toyohashi University of Technology, Japan

Zita Vale	GECAD - ISEP/IPP, Portugal
Rafael Valencia-Garcia	Departamento de Informática y Sistemas, Universidad de Murcia, Spain
Miguel A. Vega-Rodríguez	University of Extremadura, Spain
Maria João Viamonte	Instituto Superior de Engenharia do Porto, Portugal
Paulo Vieira	Instituto Politécnico da Guarda, Portugal
José Ramón Villar	University of Oviedo, Spain
Friederike Wall	Alpen-Adria-Universitaet Klagenfurt, Austria
Zhu Wang	XINGTANG Telecommunications Technology Co., Ltd., China
Li Weigang	University of Brasilia, Brazil
Bozena Wozna-Szczesniak	Institute of Mathematics and Computer Science, Jan Dlugosz University in Czestochowa, Poland
Michal Wozniak	Wroclaw University of Technology, Poland
Takuya Yoshihiro	Faculty of Systems Engineering, Wakayama University, Japan
Michifumi Yoshioka	Osaka Pref. Univ., Japan
Agnieszka Zbrzezny	Institute of Mathematics and Computer Science, Jan Dlugosz, Poland University in Czestochowa, Poland
Andrzej Zbrzezny	Institute of Mathematics and Computer Science, Jan Dlugosz University in Czestochowa, Poland
Zúquete	University of Aveiro, Portugal

Organizing Committee

Juan Manuel Corchado Rodríguez	University of Salamanca, Spain and AIR Institute, Spain
Sara Rodríguez González	University of Salamanca, Spain
Roberto Casado Vara	University of Salamanca, Spain
Fernando De la Prieta	University of Salamanca, Spain
Sonsoles Pérez Gómez	University of Salamanca, Spain
Benjamín Arias Pérez	University of Salamanca, Spain
Javier Prieto Tejedor	University of Salamanca, Spain and AIR Institute, Spain
Pablo Chamoso Santos	University of Salamanca, Spain
Amin Shokri Gazafroudi	University of Salamanca, Spain
Alfonso González Briones	University of Salamanca, Spain and AIR Institute, Spain
José Antonio Castellanos	University of Salamanca, Spain
Yeray Mezquita Martín	University of Salamanca, Spain

Enrique Goyenechea	University of Salamanca, Spain
Javier J. Martín Limorti	University of Salamanca, Spain
Alberto Rivas Camacho	University of Salamanca, Spain
Ines Sitton Candanedo	University of Salamanca, Spain
Daniel López Sánchez	University of Salamanca, Spain
Elena Hernández Nieves	University of Salamanca, Spain
Beatriz Bellido	University of Salamanca, Spain
María Alonso	University of Salamanca, Spain
Diego Valdeolmillos	University of Salamanca, Spain and AIR Institute, Spain
Sergio Marquez	University of Salamanca, Spain
Guillermo Hernández González	University of Salamanca, Spain
Mehmet Ozturk	University of Salamanca, Spain
Luis Carlos Martínez de Iturrate	University of Salamanca, Spain and AIR Institute, Spain
Ricardo S. Alonso Rincón	University of Salamanca, Spain
Javier Parra	University of Salamanca, Spain
Niloufar Shoeibi	University of Salamanca, Spain
Zakieh Alizadeh-Sani	University of Salamanca, Spain
Jesús Ángel Román Gallego	University of Salamanca, Spain
Angélica González Arrieta	University of Salamanca, Spain
José Rafael García-Bermejo Giner	University of Salamanca, Spain
Pastora Vega Cruz	University of Salamanca, Spain
Mario Sutil	University of Salamanca, Spain
Ana Belén Gil González	University of Salamanca, Spain
Ana De Luis Reboredo	University of Salamanca, Spain

Contents

Artificial Intelligence Applications

Development of a Dangerous Driving Suppression System Using Inverse Reinforcement Learning and Blockchain

Koji Hitomi[1], Kenji Matsui[1(✉)], Alberto Rivas[2],
and Juan Manuel Corchado[2]

[1] Faculty of Robotics and Design, Osaka Institute of Technology, Osaka, Japan
mlm18r20@st.oit.ac.jp, kenji.matsui@oit.ac.jp
[2] BISITE Digital Innovation Hub, University of Salamanca, Salamanca, Spain
{rivis,corchado}@usal.es

Abstract. Casualty injury rate in car accident is still high level. The number of annual traffic accident casualties in the world today is as much as 1.35 million, and those accidents are caused by reckless driving such as signal ignoring and over speed. In this research, we propose a system which can encourage drivers to make safe driving voluntary using a driving manner evaluation mechanism. Our proposed system uses both inverse reinforcement learning and block chain platform. As for the system development environment, we use a small robot car with a camera attached to the front of the car, and operate on a test course simulating a single lane road. Using the image from the camera, each state corresponding to the image is evaluated and reward value is assigned using inverse reinforcement learning. Either giving reward according to the evaluation value or creating rankings by verifying whether the driving accuracy is improved, the proposed system can make good motivation with competitive spirit.

Preliminary subjective test was performed with 9 subjects who drove a small vehicle. The test result shows positive feedback in case of both giving rewards and giving better ranking. ANOVA result shows that there is a significant difference at a significance level of 5%.

Keywords: Safe driving · Inverse reinforcement learning · Blockchain

1 Introduction

Road traffic injuries is still high level. The number of annual traffic accident casualties in the world today is as much as 1.35 million. Road traffic injuries cause considerable economic losses to individuals, their families, and to nations as a whole. These losses arise from the cost of treatment as well as lost productivity for those killed or disabled by their injuries, and for family members who need to take time off work or school to care for the injured. Road traffic crashes cost most countries 3% of their gross domestic product [1]. This is a serious problem. Causes of accidents are caused by a decline in driver's awareness of safe driving, such as signal ignoring and over speed. As one of countermeasures, an automobile that the car itself drives without driving through a

© Springer Nature Switzerland AG 2020
F. Herrera et al. (Eds.): DCAI 2019, AISC 1003, pp. 3–9, 2020.
https://doi.org/10.1007/978-3-030-23887-2_1

driver is a topic. But there is a concern that too safe driving by AI will induce dangerous driving of others. Also, practical application of automatic driving vehicles on public roads has problems in various aspects such as technology, ethics, law, and it is expected that it will take time for practical use.

1.1 ISA (Intelligent Speed Adaptation)

In excess of speed, in recent years, a function (ISA) to suppress the speed by comparing the traveling speed of the own vehicle with the regulation speed of the current location has attracted attention. ISA has a forced type in which the speed can not be increased with the control of the driver side and an advisory type in which the speed information is provided to the driver, etc. However, in recent years, movement to give rewards for safe driving has spread to multiple countries, And in fact, a high speed restraining effect is recognized [2, 3].

1.2 Inverse Reinforcement Learning (IRL)

Reinforcement learning (RL) which is one of machine learning has begun to surprise the performance in tasks such as games. In many cases, you can model the problem in a Markov decision process form and look at the best or near-optimal policy from the state, action, and reward. However, unlike states and actions where values are uniquely determined by information from sensors. it is often difficult to manually determine rewards depending on tasks. For example, there are tasks such as maintaining a safe inter-vehicle distance for a traveling task on a highway, maintaining a reasonable distance away from curbstones. Among these various elements it is necessary to assign a series of weights that accurately indicate how to determine the reward function how to trade off. This is a difficult thing. Therefore, there is a method called inverse reinforcement learning to solve the problem of deriving the reward function from the action and state of experts observed [4]. By using IRL, it is shown to find a policy to execute in the same way as experts in driving tasks [5]. RL and IRL have the relationship as shown in Fig. 1.

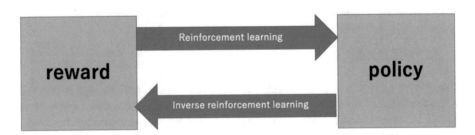

Fig. 1. Reinforcement learning and inverse reinforcement learning

2 Proposed System

In this research, safe driving is quantified using IRL with driver's driving performance data. The quantified value is taken as the evaluation input. The system gives reward according to the evaluation result, or creates a ranking to encourage competitiveness. It is aimed to improve the driving performance by adding those functions. We also verify that the safe driving has been quantified by IRL.

2.1 Architecture of IRL

In this research, the reward is the output from driving performance of an ordinary driver compared with a model learned from expert driver's driving performance. Also, the reward is used as the evaluation value of the driver. As for the IRL architecture, we use maximum entropy deep IRL [6] in this study. The maximum entropy deep IRL is a framework of exploiting fully convolutional neural networks for the reward structure approximation in IRL. The structure of the IRL is shown in Fig. 2. The image input is normalized and the size is 72 * 128 with 3 channels. Linear activation function is used in the output layer, and relu function for others.

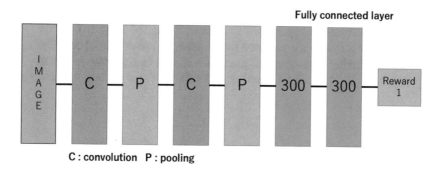

Fig. 2. Architecture of Inverse reinforcement learning

2.2 Blockchain

In this research, distribution of rewards and record of rankings are created by Smart Contract on Ethereum's test net. We distribute tokens which we created as rewards. Distributing with Smart Contract can be done without going through brokers, in order to save the cost. Another advantage is that the transaction history can be preserved semi-permanently and the system will be used with no restriction.

3 Case-Study

As for the preliminary system test, we created a pseudo elliptical one-lane road as shown in Fig. 3, which is a yellow tape and a white tape on a black mat.

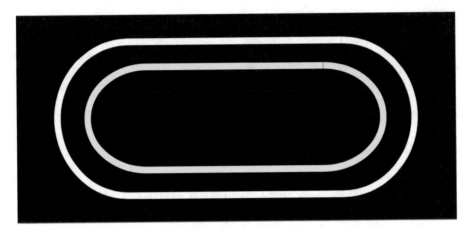

Fig. 3. One-lane road

The width of the lane is 21 cm. A small robot car is used and controlled by test subjects assuming real car driving condition. The size of the robot car is 21 cm in length and 17 cm in width. Nine Japanese adults were participated the experiment. A small wide angle camera was attached to the front of the robot car as shown in Fig. 4, and driving performance of each participant were evaluated using IRL.

The example performance was used for the learning phase with +1 for good driving and 0 for bad driving. Examples of good driving and bad driving are shown in the Fig. 5. We used 13,159 data for learning and 4,387 data for testing. Accuracy rate is 97%.

Fig. 4. Robot car

Reward +1 Reward 0

Fig. 5. Examples of learning data

The safe driving here means driving without running off the lines at both ends. The evaluation value, output by IRL, is added for each image frame. After added 70 times, the evaluation value is finalized. As for the experiment, the test subjects first practice the driving operation. Then, they perform the required operation three times.

- Subjects try to drive nomally. (Nomal)
- While testing, subjects see other driver's reward ranking list. (Ranking)
- Based on the performance of each subject, the smart contract sends tokens to the driver as a reward. (Reward)

The total system is shown in Fig. 6. After the test driving, subjects were asked whether they always acted in strong consciousness of safe driving compared with "Normal" condition.

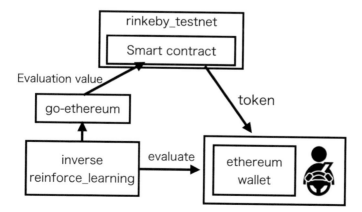

Fig. 6. Schematic diagram of the system

3.1   Results

Table 1 shows the average of the evaluation values in case of "Normal", "Ranking" and "Reward" by nine subjects. Both "Ranking" and "Reward" show 3 points higher compared with the "Normal". The result shows that there is the significant difference at the significance level of 5%. Also, from the questionnaire, all drivers answered that they had consciousness of safe driving than when there was no reward or ranking. In addition, since the evaluation value in the case of giving the reward is rising, it was confirmed that it was possible to judge the safe driving in the inverse reinforcement learning. When comparing the "Ranking" and the "Reward", there is no significant difference, the same effect can be obtained.

Table 1. Evaluation values at the first, second and third times

	Normal	Ranking	Reward
Average evaluate value (point)	36.1	39.2	39.3

3.2   Discussion

There are a couple of safe driving warning system available. Those systems are mainly using accelerometer, GPS, tachometer, etc. [7] [8]. Some of those research results show improved driving by giving advice generated by those warning system [9]. In this research, by using single wide-angle camera, it is possible to acquire more detailed warning information such as road conditions. Also together with RIL, the system can evaluate the driving manner more precisely and objectively. Furthermore, showing the ranking and reward tokens seems better way of making warning. Although our study is still very preliminary level, we were able to see one of the interesting approach using IRL and Blockchain.

4   Conclusions

In this research, we developed and tested a preliminary driving warning system using IRL and Blockchain, which gives reward mechanism according to the evaluation value. That system has a potential to generate safe driving ranking which can encourage drivers to maintain strong consciousness of safe driving. We tested the effectiveness of the proposed system using small robot car based experiment. As a result, using "ranking" and "reward" we found that there was a significant difference at the significance level of 5%. As for the effectiveness of "ranking" and "reward", there was no significant difference.

For our future work, we plan to increase the states which can be evaluated by IRL so that the proposed system can cover various kind of problems. We plan to work on the improvement aiming at incorporating various application systems such as drive recorder. By using the block chain, it is possible to save the image in a non-alterable state especially when the reward is low. That information will be quite valuable for insurance company. There will be a chance to be able to lower the insurance cost by utilizing the proposed system which could make more reliable data.

References

1. World Health Organization: Global Status Report on Road Safety (2018). https://www.who.int/news-room/fact-sheets/detail/road-traffic-injuries
2. Mazureck, U., Van Hattem, J.: Rewards for safe driving behavior: influence of following distance and speed. Transp. Res. Rec. **1980**, 31–38 (2006)
3. Lahrmann, H., Agerholm, N., Tradisauskas, N., Berthelsen, K.K., Harms, L.: Pay as you speed, ISA with incentive for not speeding: results and interpretation of speed data. Accid. Anal. Prev. **48**, 17–28 (2012)
4. Ng, A.Y., Russell, S.: Algorithms for inverse reinforcement learning. In: Proceedings of ICML (2000)
5. Ziebart, B.D., et al.: Maximum entropy inverse reinforcement learning (2008)
6. Wulfmeier, M., Ondruska, P., Posner, I.: Maximum entropy deep inverse reinforcement learning (2016)
7. Jachimczyk, B., Dziak, D., Czapla, J., Damps, P., Kulesza, W.J.: IoT on-board system for driving style assessment (2018)
8. Rubira Freixas, M.: Effects of driving style on passengers comfort: a research paper about the influence of the bus driver's driving style on public transport users. Bachelor's thesis, KTH Royal Institute of Technology, Stockholm, Sweden (2016)
9. Scania Group: Scania Support Tools for Drivers and Operators (2019). https://www.scania.com/group/en/scania-support-tools-for-drivers-and-operators

Korean License Plate Recognition System Using Combined Neural Networks

Saidrasul Usmankhujaev, Sunwoo Lee, and Jangwoo Kwon[✉]

Department of Computer Science and Engineering, Inha University,
Incheon, Republic of Korea
{u.s.saidrasul, x21999}@inha.edu, jwkwon@inha.ac.kr

Abstract. We developed a deep learning application to detect and recognize Korean cars' license plates from images. It is an advanced application that targets to provide deep learning solution that can be applied in many areas including Intelligent Transportation System, Internet of Things and Smart City. Despite, there have been many approaches and studies on license plate localization, character segmentation and recognition, there have not been highly demanded results particularly using deep neural networks. Traditional approaches on license plate detection have achieved quite a high accuracy in detection and recognition, in which mostly Optical Character Recognition (OCR) is used. Nevertheless, in this research, we developed our own method that is a combination of scene text recognition technique with Geometrical Image Transformation (GIT) to recognize number plates for combined neural networks and achieving 99.8% and 95.7% of detection and recognition accuracy respectively.

Keywords: CNN – Convolutional Neural Networks ·
YOLOv3 – You Only Look Once · RNN – Recurrent Neural Networks ·
LSTM – Long Short-Term Memory ·
CRNN – Convolutional Recurrent Neural Networks

1 Introduction

Automatic Number Plate Recognition system (ANPR) - a surveillance technology that is mostly based on optical character recognition applied on pictures to understand the vehicle registration plates' locality and then using character segmentation extract the plate data [1]. Mainly ANPR systems are like an embedded software that is implemented in road enforcement cameras.

Task of plate recognition using OCR is quite challenging problem due to the various types of plate formats and the irregular lighting conditions. Even though nowadays there are powerful ANPRs with high efficiency of recognition, the technique itself has number of limitations in processing captured images. These difficulties are low image resolution, blurry picture or object is located too far to be recognized. However, the most usual case of misrecognition is poor lighting conditions.

Our proposed method (GIT) is based on the deep image analysis and learning by transforming the image. In the first part of our work, we trained the YOLOv3 - the state-of-the-art method for real-time object detection, which has faster training

F. Herrera et al. (Eds.): DCAI 2019, AISC 1003, pp. 10–17, 2020.
https://doi.org/10.1007/978-3-030-23887-2_2

convergence and competitive mean average precision (mAP) for boundary boxes compared to other methods described in [2]. We prepared our own dataset with around 50000 CCTV images of cars taken from different angles consisting of numerous numbers of dark images where the number plate could hardly be recognized even by human eyes.

For the recognition part we used another network architecture named Convolutional Recurrent Neural Networks which is a combined neural network to identify the sequence of characters and numbers from the image [4]. To train the CRNN network we required huge number of plates' images because the learning is based on sequence learning rather than feature learning, therefore we created a synthetic license plate generator [9]. In an addition to our dataset we generated nearly 60000 plates' images and made our dataset half synthetic.

To generate plates, we used functions provided by OpenCV libraries. We found out the real size of a plate and exact position of each character on a plate. Then we cropped every single character and digit to place it on a generated background using OpenCV. Nevertheless, each type has different dimensions and different size of characters, so we considered this matter also. We generated exact size of number plate. For example, if the size of a plate is 520 × 110 mm when converting to pixel values with our screen dpi = 72 the picture becomes 1679 × 355 pixels (Fig. 1).

Fig. 1. Character position on a license plate

2   Vehicle Registration Plates of the Republic of Korea

There are several valid types of car plates which were designed by Ministry of Land, Infrastructure and Transport of South Korea. The newest and the most widely used types are being produced since a year of 2006. Particularly we chose 5 the most widely used plates and separated them according to their shape and appearance (Fig. 2).

Each type has its own specific number of unique Korean characters used in the plate. For example, for type 1, 2 and 5 there are 50 specific characters that can be considered as classes to be classified, for type 3 and 4 there are 36 characters respectively, etc. The reason for this every car plate type is specific because in some plates the city/province names were used and only 5 Korean characters used as a second symbol.

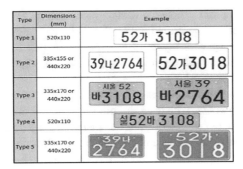

Fig. 2. Valid Korean cars' number plates

3 Related Work

3.1 CNN-Based Approach for Car Number Plate Recognition

Nguyen et al. [3] have created a small convolutional architecture to recognize Vietnamese car number plates. They created the CNN-based method and their key idea was to combine detection, segmentation, and recognition of number plates in one. Their method was reluctant to the haar-like cascade, that are digital image features used in object detection and face recognition, with the combination of the histogram of gradients (HOG) which calculates appearance of gradient orientation in various portions of an image. In this work, they applied Google's Multi-digit Number Recognition from street view to the recognition of the number plates using a single end-to-end neural network [5]. That method is the combination of three steps mentioned above (localization, segmentation, and recognition) using Convolutional Neural Networks (CNN) [6]. According to authors their proposed system is quite slow to process one image and get the predicted output. Also, they used image processing technique with the combination of machine learning algorithm, while we propose the deep learning solution of a problem.

3.2 Convolutional Recurrent Neural Networks

Shi et al. have proposed their own method of recognizing the scene text from an image by using the combination of CNN and the Recurrent Neural Networks (RNN) where they divided whole work into three components including convolutional layers, recurrent layers and transcription layers (Fig. 3) [7].

First, grayscale image is given as an input to the VGG network [8] and convolutional feature maps are obtained, afterwards these maps are fed into feature sequence and passed to the recurrent layers, where 2 layer bidirectional LSTM stack was used, to predict a label distribution for each frame, then this per-frame predictions distribution is sorted to proceed the final text output. The main idea was to identify and understand the text in the wild and create an end-to-end trainable network and experiment on standard benchmarks like Street View Text and ICDAR achieve better results than their predecessors in text recognition. In the proposed method, the component of convolutional layers is constructed having the convolutional and max-pooling layers of standard VGG network with removed fully connected layers and sequence extraction is done by taking features from convolutional and max-pooling layers. Extracted feature vectors become an input for the recurrent layers. In our approach, we changed the training parameters and by using image processing techniques we could achieve much better results.

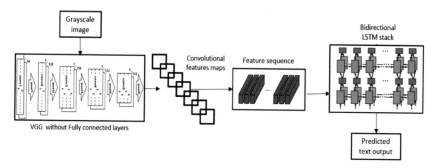

Fig. 3. CRNN model representation

4   Proposed System

In this section, we described in detail how we detected plates from images and the usage of CRNN to recognize the plates. In previous sections, we divided the job into two main parts and then we prepared about 9500 images to train YOLOv3 [2] network to identify the position of a number plate. Basically, YOLO completes several algorithms at once including localization, orientation, sizing and segmentation with image processing.

4.1   License Plate Detection

To increase overall performance of the detection model we needed to have "balance" in dataset. By "balance" we mean that different background images should be proportional. We had imbalanced dataset with having majority of CCTV images over parking ones. Therefore, we equalized the number images taken from different angle by augmenting them. As augmentation techniques we used horizontal and vertical flip and the combination of both. These simple augmentations can be applied to object detection because basically by rotating the image we just increase the number of images in the dataset.

We trained YOLOv3 to identify plates in the dark by changing the images histogram values so that only required zone is brightened to be properly extracted (Fig. 4). YOLOv3 predicts an object score and type of a plate for each bounding box using logistic regression with reasonable mAP score. However, the scene text recognition was working partially in our problem. Therefore, we tested our contribution in comparison with the original scene text recognition method proposed by Shi et al. [4].

Fig. 4. Image with enhanced histogram values of detected parts

4.2 Scene Text Recognition with Geometrical Image Transformation

First, we used 100×32 pixel gray-scale images to train and obtain the label sequence from one-line plate, where feature learning was only accordingly to y-axis. Every learned region is called perceptive field, and each perceptive field is fed to the particular unit in the LSTM stack, [10] and since we have bidirectional LSTM stack the features extracted forward and backward using Connectionist Temporal Classification (CTC) layer where the probability of a character is defined (Fig. 5) [11]. However, in double row numbers, like type 3 and 5, there is a possibility that two characters may occur in one perceptive field and the character with the highest probability is chosen and another is simply ignored. This led to a problem of partial detection. Therefore, we use geometrical image transformation technique to skew the image and make every single character not to lay in one perceptive region that is fed to LSTM. When we transform only the double row images accordingly (Fig. 6), detection accuracy significantly increased.

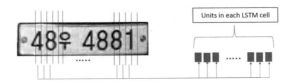

Fig. 5. Visualization of perceptive areas which are fed into LSTM. This part is like classification part because every character is considered as a class and probability is calculated to which class single character belongs to.

GIT technique doesn't change the image content but just change the pixel grid by deforming it to the destination image and can be applied only for 2D images. We used 112×32 pixel gray-scale image with 256 units in each LSTM layer that for every image there will be 256 perceptive regions and each region is stacked to each unit where forward-backward probability is calculated. Also, we indicated the sequence length which is the length of all numbers and characters of a license plate.

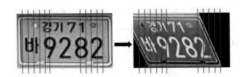

Fig. 6. Visualization of perceptive areas with transformed image

5 Experiment Results and Evaluation

5.1 Datasets

For experiments we used half real, half synthetic data. First, we divided the dataset into two parts by randomly selecting training and test images as 85% to 15% respectively. We acquire YOLOv3 with 9500 different quality and background view images. However, for training the CRNN model we didn't require to have image of a car, but only license plate Using YOLO we cropped a license plate from the car on image and fed that plate to CRNN. The detection accuracy using YOLO was quite impressive with 99.8%.

5.2 Networks Training

All trainings were done on Windows machine having CPU Intel Core i5-7600, 3.50 GHz and GPU Nvidia GeForce GTX 1080Ti, with 16 GB RAM. To train YOLOv3 network we chose YOLOv3-416 which converts images to 416×416 pixels, with learning rate $\alpha = 1e-3$, batch size = 64 on frozen layers and batch size = 8 for unfrozen ones, and we used *Adam* optimizer with $\alpha = 1e-3$ for frozen layers and $\alpha = 1e-4$ after unfreezing the layers [2]. Training in YOLO is more time consuming and took about 25 h to a network to get its convergence.

For training CRNN model we prepared only images of license plates. Originally Shi et al. [4] trained 8 million synthetic images from Synth dataset for text recognition, however we couldn't have such big dataset and we prepared only about 110.000 images because we mostly concerned about real world data recognition and manually separate whole dataset according to plate type (Table 1).

Table 1. Number of images in our own dataset

Type #	Real images	Generated images
1	20626	11378
2	3355	11882
3	6013	25196
4	10392	10000
5	2810	10277

The CRNN convergence is faster and it processes images much faster than YOLOv3 due to the smaller image size of 112×32 assigned for VGG. Training time usually varies from 3 to 6 h. For training the model we used following hyperparameters: initial learning rate $\alpha = 0.15$ with 20000 iterations and learning rate decay to $\alpha = 1e-3$ every 5000 steps, with 128 and 64 for training and validation batch sizes. For example, if we have 32768 images one epoch will cost 256 iterations. We used *Adadelta* [12] because it has faster convergence than traditional *Momentum* optimizer, with learning rate $\alpha = 0.97$.

5.3   Results

The project is done on Python language with usage of TensorFlow for deep learning calculations. For both networks we used TensorFlow backend but for YOLOv3 we also used Keras high level API [13]. In first method, we used original CRNN network and prepared model trained on real + synthetic images without any image enhancement techniques. For testing we have taken 5552 real world images and run prepared model. Then, we created separated models for each type. Then, we applied the color inversion method to separate models and finally proposed our contribution with GIT. Every method was run on the same randomly selected real-world images. Table 2 describes the number plate recognition accuracies of different methods.

Table 2. Recognition accuracy (%) comparison table.

Method	Type 1	Type 2	Type 3	Type 4	Type 5	Average
CRNN	95.6	95.0	66.2	92.9	79.1	85.7
CRNN + color inversion	99.7	99.4	76.9	99.6	76.5	90.4
CRNN + GIT + color inversion	**99.7**	**99.4**	**90.8**	**99.6**	**88.6**	**95.6**

(*Remark:* GIT is applied only to double row images as: type 3 and type 5, color inversion wasn't applied for type 5 because the digits and character are white)

As we can see that the color inversion increases the overall accuracy of recognition. The reason is that when we extract plate from YOLO the image has poor quality and since, there were many dark plates, inverting the color simply flips the RGB values and emphasizes better sequence learning.

6   Conclusion and Future Work

Even though the text recognition is one of the complicated tasks, this proposed project using deep learning identifies the numbers and characters from image with high accuracy. This research shows that deep learning can have competitive efficiency with traditional ANPR systems and much more, the main advantage of Deep Learning is that it's significantly cheaper to be implemented and there will be no need for expensive surveillance cameras because every single camera can work as surveillance one. Also, the usage of neural networks together with traditional approaches can significantly increase the performance of any recognition system because we practically showed that even poorly visible images can be understood by neural networks. Moreover, this project is a new tool for ITS projects and the true implementation for Smart City concept, which clearly points to the worldwide fourth industrial revolution. However, we still have works to be done further. We only consider the 5 license plate types, despite there are more types, we didn't consider very old and rare types, and because our dataset was taken from real world cars and the ratio of those plates are 1/200. These types can be considered as a next step in our research.

Acknowledgment. This research was supported by the MSIT (Ministry of Science and ICT), Korea, under the ITRC (Information Technology Research Center) support program (IITP-2017-0-01642 and IITP-2018-2014-1-00729) supervised by the IITP (Institute for Information & Communications Technology Promotion).

References

1. Qadri, M.T., Asif, M.: Automatic number plate recognition system for vehicle identification using optical character recognition. In: International Conference on Education Technology and Computer (2009)
2. Redmon, J., Farhadi, A.: YOLOv3: an incremental improvement. arXiv:1804.02767v1 (2018)
3. Nguyen, T., Nguyen, D.: A new convolutional architecture for Vietnamese car plate recognition. In: 10th International Conference on Knowledge and Systems Engineering, KSE (2018)
4. Shi, B., Bai, X., Yao, C.: An end-to-end trainable neural network for image-based sequence recognition and its application to scene text recognition. IEEE Trans. Pattern Anal. Mach. Intell. (2017)
5. Goodfellow, I.J., Bulatov, J., Ibarz, J., Arnoud, S., Shet, V.: Multi-digit number recognition from street view imagery using deep convolutional neural networks. In: International Conference on Learning Representations, ICLR arXiv:1312.6082 (2014)
6. Kyaw, N.N., Sinha, G.R., Mon, K.L.: License plate recognition of Myanmar vehicle number plates a critical review. In: 2018 IEEE 7th Global Conference on Consumer Electronics, GCCE (2018)
7. Mikolov, T., Karafiat, M., Burget, L., Cernocky, J., Khudanpur, S.: Recurrent neural network-based language model. In: 11th Annual Conference of the International Speech Communication Association (2010)
8. Simonyan, K., Zisserman, A.: Very deep convolutional networks for large-scale image recognition. In: International Conference on Learning Representations (2015)
9. Github. https://github.com/Usmankhujaev/KoreanCarPlateGenerator. Accessed 26 Mar 2019
10. Hochreiter, S., Schmidhuber, J.: Long short-term memory. Neural Comput. **9**(8), 1735–1780 (1997)
11. Graves, A., Fernandez, S., Gomez, F., Schmidhuber, J.: Connectionist temporal classification: labelling unsegmented sequence data with recurrent neural networks. In: Proceedings of the Twenty-Third International Conference on Machine Learning, ICML (2006)
12. Zeiler, M.D.: ADADELTA: an adaptive learning rate method. arXiv:1212.5701 (2012)
13. Chollet F.: Keras. GitHub (2015)

A Comparative Study of The Corpus for Story Creation System

Haruka Takahashi$^{(\boxtimes)}$, Miki Ueno, and Hitoshi Isahara

Toyohashi University of Technology, 1-1 Hibarigaoka, Tempaku-cho,
Toyohashi, Aichi, Japan
takahashi@lang.cs.tut.ac.jp

Abstract. In recent years, there are lots of researches to generate creations such as comics and novels by computing methods. Especially, the orders of sentences are very important for the quality of stories. Thus, the aim of the researches is to construct story creation system in order to consider the flow of the story based on the sentence similarity. We suggest the possibility to create a new story by selecting sentences from exiting novels based on similarity. We indicated that the length of sentences and numbers of proper nouns tend to affect the naturalness of the created story.

Keywords: Story creation system · Supporting contents creation ·
Distributed representation ·
Contents creators and artificial intelligence · Searching similar sentences

1 Introduction

In recent years, learning using a neural network due to improvement of computer performance is thriving [1]. From these researches, it is considered that the computer can also assist human beings in creative activities.

The story is one of the most important components of the creation and is related to the evaluation and identity. Therefore, it is a challenging task to process the method of creating a story on a computer.

In this research, we will construct an automatic story creation system to analyze the story composition that the reader does not feel unnatural, and to investigate whether it can be applied to the creation of new works by changing the order of the sentences.

2 Related Study

2.1 Supporting Contents Creation

As a previous research, there is a creation support System to manage the Story structure based on template sets and graph [2]. In this research, the aim is

© Springer Nature Switzerland AG 2020
F. Herrera et al. (Eds.): DCAI 2019, AISC 1003, pp. 18–27, 2020.
https://doi.org/10.1007/978-3-030-23887-2_3

Table 1. Parameters of Doc2Vec

Method	Dimension size	Min count	Epoch
DM-PV	100	1	100

DM-PV: Distributed Memory Paragraph Vector [6]

to create a creation support system, analyze the creation process and support creation from input to the system. From the categories prepared beforehand, the creator can select a suitable one for the story to be devised this time and create a story plot in a form to fill the form on the software. However, the creation of the story is by the creator, and the computer does not automatically create it.

2.2 Story Creation

As a previous research of automatic generation of sentences, there is chat response generation based on semantic prediction using distributed representation of words [3]. In this research, they use Recurrent Neural Network (RNN) to learn conversation sentences on Twitter and create appropriate response sentences for input of a certain conversation sentence. Unlike conventional one-hot expressions, they use distributed expressions of words as inputs. This research is intended for conversational sentences so that output sentence from one input sentence and does not correspond to output of multiple sentences such as stories.

As research to generate grammatically correct sentences, there is the method of automatic generation of very short stories by imitation and substitution [4].

3 Construction of Story Creation System

In this research, we construct a story generation system based on sentence similarity by using distributed representation of sentences.

3.1 Corpus Creation

Three pieces of data are used for the corpus. The data used is shown below.

1. Manga 109's 4-scene comics' serif data (5145 sentences) [7]
2. Synopsis data of Aozora Bunko (7439 sentences)
3. Text data gathered from the novel posting site on the Internet (383494 sentences)

Due to large number of sentences, we used the sentences collected from the web pages posted novels. The sentences were divided by a punctuation mark "." and were used as a set of one-line sentences.

3.2   Paragraph Vector Model

We create a model of paragraph vector. Paragraph vector is a technique to convert a sentence into a distributed representation [5]. By using vector representation, the similarity of sentences can be calculated by the cosine distance and so on.

We extract words from sentences used for training paragraph vector. We ignore parts of speech: particles, auxiliary verbs, symbols and emoticons by morphological analysis with the library MeCab based on NAIST Japanese Dictionary[1]. The parts of speech used for the training are noun, verb, adjective, adverb, adnominal.

Paragraph vector were created by Doc2Vec included in Gensim[2]which is the popular library made by Python. Table 1 shows the parameters of Doc2Vec.

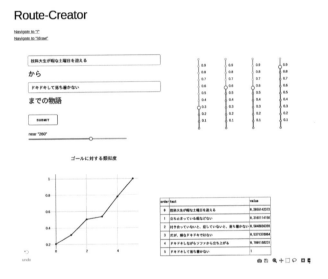

Fig. 1. The user interface of the system

3.3   The Method of Route Creation

Using the created Paragraph Vector model, a procedure of creating a story be described. The input of the system is the "start sentence r_s" and "goal sentence r_g" of the story. The transition route of the story is created by the following procedure so that the similarity is higher from the start to the goal.

(1) Prepare an empty queue $\mathbf{R}_s$. The sentence that exists in the model at the position closest to the sentence of the goal is regarded as the temporary goal sentence r_g'. Add r_s to $\mathbf{R}_s$.

---

[1] NAIST Japanese Dictionary, http://naist-jdic.osdn.jp/.
[2] Gensim, https://radimrehurek.com/gensim/.

Table 2. An example of the created story

	Sentence	Similarity
0	TUT student get free Saturday	0.206
1	There is no free time to stop	0.316
2	I feel restless if I am not dating or I do not fall in love	0.505
3	However, it is not a bad throbbing	0.537
4	Standing up from the sofa while throbbing	0.780
5	My heart is throbbing, I am restless	1.00

The original sentences are written in Japanese.

Table 3. The results of the questionnaire

Pattern	Number of neighborhoods	Number of sentences	Felt natural	Felt almost natural	Felt somewhat unnatural	Felt unnatural
1	50	9	0	1	2	1
	100	7	0	1	2	1
	300	7	0	2	2	0
2	50	29	0	0	1	3
	100	9	0	1	2	1
	300	11	0	0	4	0
3	50	7	0	0	3	1
	100	6	1	1	2	0
	300	6	0	0	2	2
4	50	7	1	2	1	0
	100	9	1	2	1	0
	300	4	2	1	1	0

(2) Nearly n sentences were acquired for the sentence at the end of $\mathbf{R}_s$ and sorted in descending order of similarity with r'_g. Create array $\mathbf{A}_s = \{a_1, a_2, ..., a_n\}$.

(3) A sentence a_i that is not existing in $\mathbf{R}_s$ and has the smallest index number in a sentence included in $\mathbf{A}_s$ is added to $\mathbf{R}_s$.

(4) If r'_g exists in $\mathbf{A}_s$, r'_g and r_g are added to $\mathbf{R}_s$, and route creation is completed. If it does not completed, go back to (2) and repeat the process.

The created $\mathbf{R}_s$ is the story.

3.4   User Interface Implementation

This research is aimed at creative support and assuming that users will use it easily in the future, we implemented this story creation system as a WEB application. This application is made by Python and is based on the library

Dash by plotly[3] for creating WEB application. Figure 1 shows the user interface of the system. The upper left corner of the Fig. 1 is a text box for user's inputting.

The similarity adjustment function on the upperright of the Fig. 1 determined the duration of the similarity between two continuous sentences in the story. The created story is displayed in the lower right. In the lower left, the trajectory of the transition of the similarity is displayed as a graph.

3.5   Examples of Generated Sentences

Table 2 shows an example of the generated story. In the generation result, the number of sentences was fixed to 6 sentences including the start and goal sentences, and the similarity was adjusted like the slider at the upper right of Fig. 1.

4   Experiment 1

4.1   Questionnaire Evaluation

First, we conducted a questionnaire to evaluate this system. We input four patterns with different starting and ending pairs, changed the number of neighborhoods to three (50, 100, 300) kinds and created stories respectively from the input sentences. We asked four users to evaluate those stories in four levels (felt natural, felt almost natural, felt somewhat unnatural, felt unnatural). Table 3 shows the results of the questionnaire.

4.2   The Comparative Result by System and Random Generation

We asked four users to compare with 8 patterns of stories created by random selection and stories created using the system. We created five continuous sentences and 10 continuous sentences by a method using the system and a random method from corpus respectively. These five sentences and 10 sentences do not include start and goal sentences. Seven sentences and 12 sentences with input added are regarded as stories. Next we asked users choose a more natural sentence out of the two sentences. Pairs of the same input and the same sentence number are used for comparison.

We fixed the nearest neighbor number to 300 because the number of users who felt unnatural was the least under the condition in the above survey. Figures 2 and 3 show the slider setting of the similarity adjustment function in case of five sentences and 10 sentences generated by the system. The slider of the similarity adjustment was fixed to prevent the intentional accents in the story.

Table 4 shows the comparison results. The numbers in Table 4 show the number of people that selected it amongst the four users.

---

[3] Dash by plotly, https://plot.ly/products/dash/.

Fig. 2. Similarity adjustment slider for 5 sentences

Fig. 3. Similarity adjustment slider for 10 sentences

Table 4. The result of comparing system and random

Pattern	5 sentences		10 sentences	
	System	Random	System	Random
1	3	1	4	0
2	3	1	3	1
3	4	0	4	0
4	4	0	3	1
5	3	1	4	0
6	3	1	4	0
7	4	0	4	0
8	2	2	4	0

4.3 Consideration

The result suggested that increasing the number of neighborhoods, the number of people who feel unnatural decrease.

In comparison with random generation, the system showed the effectiveness in both cases of five sentences and 10 sentences. However, users sometimes regarded natural story even by random generation, when stories were composed of five sentences. On the other hand, in the case of 10 sentences, four users answered that the system is more natural in almost input patterns. This system selected sentences using similarity, the number of proper nouns in a whole story is hard to increase even if the number of sentences increases.

In the case that story consisted more than five sentences, this system tends to create stories naturally compared to the method of creating stories by selecting sentences randomly. The result indicates that the length of sentences and the numbers of sentences affect the naturalness of stories. As the number of words increases, the number of proper nouns in the whole story also tends to increase. Increase of proper nouns is likely to cause inconsistency in scenes, times, or world view in the story. The improvement can be by converting proper nouns to pronouns, reducing sentences, and restricting the length of sentences. In addition, it is conceivable that the average of similarity of similar sentences acquired by corpus size changes greatly. Considering numbers of neighborhood appropriately according to corpus size is needed to sufficient scene transition.

Table 5. Changes in preprocessing

Target	Corpus in Sect. 3.1	Corpus in Sect. 5
Space	Deleted	As the end of the sentence, divide the sentence there
Japanese-style quotation marks	Replace with comma	As the end of the sentence, divide the sentence there
Double-byte symbols		Deleted except exclamation mark, question mark and double square brackets

In the case of the models have the large number of extremely similar sentences, the system tend not to scene transition with selecting similar sentences. We assume that decrease the number of extremely similar sentences at the time of model creation or temporarily increase the neighborhood n are needed to solve the problem.

5   Change of Corpus Preprocessing

In the experiment 1, the similarity of long sentences were inappropriately calculated sometimes.

Therefore, we created a corpus in a way different from experiment 1. Table 5 shows the changes in the processing of sentences. Works of the Japanese novel posting site on the Internet is written in a free format, and the appearance number of the double-byte symbols is large.

There are cases where punctuation points are not used even at the end of the sentence, accordingly there are cases where blank space is used as the break of sentences. In the previous corpus, we placed the start and end brackets together and made it a sentence. However, we decided to independently extract the sentences in the brackets.

The number of sentences as training data amount increased from 38,3494 to 66,3066 by this process.

6   Experiment 2

The result of experiment 1 shows that the long sentence tends to prevent generating natural story. We conduct a comparative experiment with a questionnaire survey of Sect. 4.1 to confirm the improvement by a new corpus as shown in Sect. 5. Comparative experiment about random generation is not addressed because the effectiveness of the our proposed system was already shown in Sect. 4.2.

6.1 Experimental Method

We do experiment of story generation using corpus modified as shown in Sect. 5. With the same parameters as Sect. 3.5, a story was also generated for models after changing pre-processing.

The number of sentences is not fixed with default similarity adjustment and same pair of input sentences as same as experiment 1.

6.2 Result and Consideration

Table 6 shows an example of the created story. Table 7 shows the results of the questionnaire.

In this experiment, the number of people who answered "felt somewhat unnatural" or "felt unnatural" decreased slightly. With regard to the point that "long sentences influence naturalness" mentioned in experiment 1, the reason of the improvement is that the sentences of the new corpus were divided into smaller ones and the number of long sentences decreased. By using the distributed representation of paragraph vector, the number of extremely long sentences decreased. Therefore, that the system can accurately calculate the similarity by the new corpus.

However, in pattern 4 of the Table 7, when using the new corpus, the number of users who answered "felt natural" was less than the Table 3. The result for the input of result 4, although there were several long sentences in the result of Sect. 4.1, there are few proper nouns and numerous pronouns such as "me" and "her" are used. It seems that users felt the story naturally because they comprehended the meaning by themselves. On the other hand, in the result of the Table 7, it became the story that repeats the same expression many times depending on the effect of increasing the number of sentences. In the current method of acquiring the top n similar sentences, we found that using the new corpus can cause the problem of multiple occurrences of the same meaning sentences described in Sect. 4.3.

Based on these facts, we need to make it difficult to be affected by the distance and number of sentences nearby. For that reason, we analyze the distance of the vector according to the scene change of the story and it is necessary to improve the method of selected sentences.

In addition, using the paragraph vector, the meaning of the sentence is sensitive to the word appearing in the sentence. As shown in the Tables 2 and 6, the same word appears in the continuous sentences. The algorithm of paragraph vector can be related to the problem. Because "not" is not taken into consideration when learning models, the distance of sentences with similar words are regarded close despite the sentiment of sentences. Thus, we should consider combining different methods for calculation of similarity.

Table 6. An example of the created story using modified corpus

	Sentence	Similarity
0	TUT student get free Saturday	−0.013
1	There is no free time to stop	0.338
2	[Character name], do you feel restless?	0.541
3	It's restless in my heart	0.568
4	I wonder what, I feel restless	0.618
5	My heart is throbbing, I am restless	1.00

The original sentences are written in Japanese.

Table 7. The results of the questionnaire using modified corpus

Pattern	Number of neighborhoods	Number of sentences	Felt natural	Felt almost natural	Felt somewhat unnatural	Felt unnatural
1	50	7	0	2	2	0
	100	6	0	3	1	0
	300	5	0	0	3	1
2	50	10	0	0	2	2
	100	10	0	3	1	0
	300	7	2	1	1	0
3	50	10	0	1	1	2
	100	6	0	2	2	0
	300	6	1	1	2	0
4	50	10	0	0	2	2
	100	10	0	0	2	2
	300	5	0	2	1	1

7   Conclusion

In this research, we created a story creation system for supporting contents creation. We suggested the possibility to create a story that people perceive as natural by re-ordering sentences based on similarity.

We mainly found the following,

- Users tend to regard the story as unnatural when there are many or long sentences with a large number of nouns.
- It is necessary to improve the method of calculating sentence similarity by converting proper nouns to pronouns.
- Removing symbols and changing the division of sentences improve the result of the created story.

Sentences that do not contain the same word but have similar meanings are closer to each other. Even that the words appearing in the sentence are similar, the meaning of sentences are opposite sometimes. It is desirable to create a model than can manage such issues. In addition, since the number of similar sentences contained in the corpus greatly affects the distance of sentences in the model, it is necessary to appropriately adjust the distance.

Acknowledgment. A part of the work is supported by JST, ACT-IJSPS (Grant number: JPMJPR17U4) and JSPS KAKENHI (Grant number: JP17K17809).

References

1. Ueno, M.: Creators and artificial intelligence: enabling collaboration with creative processes and meta-data for four-scene comic story dataset. In: The 32nd Annual Conference of the Japanese Society for Artificial Intelligence, 4Pin1-16 (2018)
2. Katsui, T., Ueno, M., Isahara, H.: A creation support system to manage the story structure based on template sets and graph. In: The 31st Annual Conference of the Japanese Society for Artificial Intelligence, 4F1-3in2 (2017)
3. Furumai, K., Takiguchi, T., Ariki, Y.: Chat response generation based on semantic prediction using distributed representations of words. In: The Acoustical Society of Japan, ROMBUNNO.2-Q-12 (2018)
4. Ogata, K., Sato, S., Komatani, K.: Automatic generation of very short stories by imitation and substitution. In: The 28th Annual Conference of the Japanese Society for Artificial Intelligence, 1C3-OS-14b-2 (2014)
5. Le, Q., Mikolov, T.: Distributed representations of sentences and documents. Google Inc., Mountain View (2014)
6. Dai, A.M., Olah, C., Le, Q.V.: Document embedding with paragraph vectors. arXiv:1507.07998 (2015)
7. Matsui, Y., Ito, K., Aramaki, Y., Fujimoto, A., Ogawa, T., Yamasaki, T., Aizawa, K.: Sketch-based Manga retrieval using Manga109 dataset. Multimedia Tools Appl. **76**, 21811–21838 (2017)

A Short Survey on Chatbot Technology: Failure in Raising the State of the Art

Francisco Supino Marcondes[(✉)], José João Almeida, and Paulo Novais

Centro ALGORITMI, University of Minho, Braga, Portugal
francisco.marcondes@algoritmi.uminho.pt, {jj,pjon}@di.uminho.pt
http://algoritmi.uminho.pt/

Abstract. This short survey aimed initially to explore the existing state of the art for the application of chatbot on fighting (and not on spreading) of fake-news. It was then realized that there is not common to use chatbots with this "virtuous" purpose. Therefore, after two surveys and a meta-analysis, the topic had to be withdrawn since there were no survey results to discuss besides the absence of results. The survey result raised then a need to realize how chatbots are being currently used, designed and their primary sources. The result was once again confusing since, on the sample: (1) no significant concentration of usage could be found; (2) no widely adopted design strategies were identified, and (3) no significant crosscutting references to be considered as primary sources. Certainly, this can be due to a biased sample but may also be a symptom of a methodological issue on the chatbot researches. If the second possibility is proved to be right it means that chatbot research is still on a pre-paradigm stage according to Kuhn's conception. For this paper, there were performed 4 surveys with a total sample of 50 papers mostly from the last 3 years.

Keywords: Chatbot research · Chatbot state of the art · Fake-news

1 Introduction

The commercial use and researches on chatbots are increasing in the last years [17]. However, these efforts are accused of lacking theoretical foundations, being mostly based on heuristic methods [19].

This paper has then a three-fold objective. The first one is to present the survey performed on the use of chatbots for hunting fake-news; the second one is to present the survey performed on the usage, design, and foundations of chatbots; and the third one is to present a preliminary attempt to justify and frame the chatbots supposed theoretical gap based on the gathered results. Therefore, this paper follows an inductive reasoning organization. It starts by surveying a specific case, then surveys to a broader scope, that yields to some conclusions. The main conclusion is a suggestion that, based on the collected sample, chatbot knowledge may be still on the *pre-paradigm* stage *cf.* [15] as a justification and framing for the supposed theoretical gap.

© Springer Nature Switzerland AG 2020
F. Herrera et al. (Eds.): DCAI 2019, AISC 1003, pp. 28–36, 2020.
https://doi.org/10.1007/978-3-030-23887-2_4

Brief Exposition About Fake-News. The fake-news problem had become a real issue on last years as it started to be used with malicious intent through social media [21]. Its boost coincides with the sufficient development of social media chatbots also called as "social-bot" [6] and a properer understanding of the deception structure on social-media [2]. It shall be noticed that fake-news is something different from SPAM and Trolls because, even if the initial spread of a fake-news starts from a "bot-net", it tends to gain own life as actual humans start to reproduce its contents in their social medias [14].

Methodology. This paper follows the "canonical" approach for systematic reviews as described in [12] for the first survey on *the use of chatbots for fighting against fake-news* and for the survey on *chatbot applications and design approaches.* In short, it starts with an informal search in order to achieve a refined query and from the result-set select the most relevant literature by reading their title and abstracts. Among these selected papers, the key concepts are raised and discussed in relation to each other and the crosscutting references are presented.

2   The Survey Report

Following the procedure described in the Methodology (Sect. 1) the refined query with the key terms and their usual variations resulted in (as a highlight, at first the search was limited to the last three years but the retrieved amount was little, therefore this constraint was removed):

```
TITLE-ABS-KEY((chat* OR social OR conversation* OR dialogue) W/O (system OR
agent OR bot) AND (fake-news OR fake news OR misinformation))
```

This query was submitted to Scopus that retrieved 09 results. A rough classification for these papers by the reading of their title, abstract and diagonal reading of their body resulted in: social-behaviour (02); chatbot detection (04); editorial (02); and no-access (01). The main findings for this search were (1) since none of these papers addresses exactly the proposed subject and also according to [23], the use of chatbot for fighting against fake-news is, at least, not a common choice, (2) the main approach used to fight against fake-news is mostly based on chatbot detection (as it is done with SPAM and Trolls).

Since no paper matched the inclusion criteria the whole sample was rejected and a second approach called "forward snowballing" [12] was performed. The paper with the same research direction selected was [28]. That paper is similar to the proposed on this as it aims to fight fake-news by the spreading of fact-checks, it is different in the sense that it aims to enhance the action of the human "guardians" (people devoted to the spread of fact-checks). That paper reinforces the perception by the absence of citations about chatbot use for the proposed end and by stating that there is a lack of papers discussing on how to handle fake-news once they are debunked [28].

Anyway, the papers following this same research direction found on that papers references were [8,9]. The "forward snowballing" was performed through Google-scholar based on these 03 papers resulting in a sample of 14 papers

(03 bases + 11 retrieved). The same procedure used in the previous search was performed and once again it was not possible to find any paper matching the inclusion criteria. The main finding for this second search was a confirmation of the prior search: (1) chatbots are not being widely explored as tools to fighting fake-news [23, 28]; (2) fake-news are mostly being handled as SPAM and Trolls probably by ignoring that fake-news gain own life.

Once again, the whole sample was rejected but as a pattern appears to show up, a third approach called "meta-analysis" was performed through the query:

```
%bot% AND (fake-news OR misinformation OR rumour) AND (survey OR review)
```

This time the query was submitted to the ACM Digital Library narrowing the search to the computer science domain. The query retrieved 09 papers and by applying the same selection method 02 papers matched the inclusion criteria (survey papers relating bots to fake-news). To these papers, it was added another 02 from the previous searches resulting in a sample with 04 survey papers. More than corroborate to the results of the previous searches they provided a possible explanation to them. Therefore, the key-find for this search is to realize that chatbots within the fake-news domain are usually seen as malicious [26, 29, 33, 34]. In addition, since most fake-news fighting approaches involving chatbots aims to shut down these programs [26, 29, 33, 34], conceive a "virtuous" usage for them maybe not so straightforward. Nevertheless, it was not found any comment opposing to the usage of chatbots on the fight against fake-news.

It was then realized that the proposed path was open for exploration. Therefore a need for a broader survey searching for inspiration and state of the art design approaches was raised. The query

```
(%bot% AND "natural language" AND conversation)
```

submitted to the ACM Digital Library, retrieved 18 papers. The sample scattered through a wide range of use, including IoT devices controlled by chatbots, domain expert chatbots, APIs, *etc.* The same classification procedure by reading title and abstracts was performed removing 04 papers from the sample. The inclusion criteria was papers describing chatbot usage and the employed design.

Based on this 14 paper sample the first examination was to search for the base papers for the fields knowledge. This aimed to include them in the sample, eventually perform a forward or backward "snowballing", or simply use them to lay the foundations and discuss the key concepts. The union set of references for these papers resulted in 285 elements; these elements were standardized and listed. From this sample, only 02 papers of 285 were cited more than one time:

Citing papers	Citation
[4, 13, 16, 25]	WEIZENBAUM, J. Eliza—a computer program for the study of natural language communication between man and machine. *CACM* 9, 1 (1966)
[13, 20]	GRAF, B., KRÜGER, M., MÜLLER, F., RUHLAND, A., AND ZECH, A. Nombot: simplify food tracking. In: *Proceedings of the 14th International Conference on Mobile and Ubiquitous Multimedia* (2015), ACM

During the analysis, it was noticed in addition to the crosscutting papers, the existence of crosscutting *authors*. This is justifiable by the current practice of publishing partial research results resulting in the publication of several papers within the same theme by the same author. Therefore it cannot be expected that all papers get in touch with the exact same papers but they shall end to find the same author by his or her research work as a whole.

The highly cited authors in this sample are McTEAR, M., WEIZENBAUM, J., and WARD, W. This suits well since these three authors are distinguished researches in the field. However, it does not suit well the fact that these authors had appeared only on 25% of the sample (note that Turing, the creator of chatbot conception, did not arise in this list). After considering several hypotheses, possible the simpler explanation for this phenomenon is that there are not yet widely accepted theoretical foundations for chatbot technology (a weak claim).

Proceeding with the sample examination, On the 14 papers selected there are: 01 survey [13]; 01 discussion [3]; 03 books [7,10,31]; and 09 design approaches [4,16,18,20,24,25,27,30,32]. These 09 papers are then listed:

Ref	Chatbot usage	Design strategy
[4]	Individual chatbot creation	Parses the natural language interactions and sets by the example the person-specific preferences, intents, behaviors, and phrases
[16]	Automatic generation of dialogues	Provide an "wizard" similar to market solutions for building task-oriented chatbot and find the best-suited API to address the task
[18]	To keep the engagement of dialogue	Provides a hybrid approach (task and chit-chat) aiming to avoid a command-reply chatbot become a natural language command-shell
[20]	Employment of chatbots for education	Given an instructional design module with questions, to use it for learning assessment through a chat-based messaging application
[24]	Control IoT devices	Parses user-defined natural language and IF .. THEN rules into a conversation (command-reply)
[25]	Creation of domain expert chatbots	Parses a troubleshot operation into a name-entity annotated digraph, and then match it to a universal customer support dialogue skeleton
[27]	Test chatbots	Builds a supervised corpus of interactions to feed a 'seq2seq' chatbot that will be used as a testing tool for the chatbot under development
[30]	Chatbot for sales support	Models the sale experience using neural networks to classify the intent of a user during buying
[32]	Parses natural language as an API's commands	Given a phrase, identify the best-suited API to achieve the intended goal, something similar to service-oriented programming

Historically, chatbots have been used to perform chitchat conversations or dialogue based tasks as sales support. According to this sample, most of the chatbots applications are task-oriented. There are several reasons for it, for instance, it is less complex to model a fair chatbot towards a closed domain than to an open one. Another one is that task chatbots easily raise funding than chitchat ones. However, this sample also suggests that currently, it is being searched for hybrid models as it is being realized that a task-oriented chatbot without chitchat becomes a natural language based command shell. Also, that chitchat is important to hold the engagement during the dialogue enhancing the task accomplishment.

It can be said that a hinder issue for chatbot design is the corpus needed to support the dialogue. Both CBR and seq2seq approaches require a large amount of data for the corpus in order to function properly. Therefore, the sample also suggested that there are efforts directed towards this issue. Most of the efforts are directed to the automatic generation of the corpus based on an example sentence or dialogue and also by an automatic gathering of information from unstructured data or APIs according to some natural language processing feature.

The sample also shows that there also some efforts employed towards personalizing. This may relate to the data-science conception of using a huge amount of data about an actor to improve his or her experience.

There is still another issue to be considered. Apparently, there is low concordance among the field specialists about the state of the art. For instance, there are authors as [22] stating that the chatbot technology evolved in the last years, however, there also authors as [5] that understands this in the opposite way. The same is valid for chatbot design, there are authors as [1] suggesting that there are no common approaches in use while authors as [13] believes that there is so. The difficulty here appears to be on how the word *chatbot* is being employed. If the word is referring to the whole chatbot-stack (that includes natural language processing features, knowledge-bases, *etc.*) it can be said that there was tangible evolution. However, if the word is used as a synonym for the dialogue manager *cf.* [17] the same cannot be said since most of used techniques are those presented in [11]. This shows that there may be not also agreement on the terminology.

Finally, the landscape emerged for this survey was a perception that there is not yet a solid theoretical ground supporting the chatbot research and the initiatives scatter with each aiming to distinct targets. Certainly there are common *targets* as already shown but there no noticeable concentration of *usage* and *design*. This may be a symptom of the Kuhn's pre-paradigm stage that shall be further examined on further researches.

3   Conclusions

Conclusion for the First Objective. The first objective was to perform a survey on the use of chatbots (or social-bots) directed to hunting fake-news spread messages. The need for a solution as such is due to the fact that after some extent and according to how a fake-news is assembled, it may gain own life as it starts to be spread by the human actors.

Through the presented surveys, it was realized that, within the fake-news domain, chatbots are mostly considered to have "malicious" intent and therefore a "virtuous" usage ends not to be straightforward. To support this claim, on the fifty papers consulted on the surveys, it was not found any mentions to a "virtuous" usage as such. For a quote of that also points out to a gap on this subject: *"there is no work about combating fake news once it has been debunked"* [28].

Conclusion for the Second Objective. Once it was understood that chatbots are not being used to "hunt" fake-news it was decided to perform a broader survey. This new survey aimed to identify how chatbots are being used and seeking for a reference model to be used. Through the survey, it was not possible to identify any distinctive concentration or pattern for chatbot usage nor design. The only concentration is within the dialogue manager that did not evolved from [11]. It was not even possible to find a set of foundation references that crosscuts the sample. It was not performed a meta-analysis for this second survey question, however, there are some survey papers on the sample that suggests the broad direction usually taken. It may be suggested that current survey papers focus mostly on a historical perspective, and on examining the features of each chatbot implementation and their technological specificities. For a quote of a paper with a similar perception, *"the techniques of chatbot design are still a matter for debate and no common approach has yet been identified"* [1].

Conclusion for the Third Objective. Considering these two results and verifying that other surveys draw to the same conclusion a deepen summing-up is required even as a weak claim. Examining this situation through the lens of philosophy of the science it relates to Kuhn's theory on the evolution of scientific knowledge. More specifically it relates to the pre-paradigm stage where there is no theoretical consensus whose researches and research methods scatters from a large range of concurrent and overlapping conceptions. By one side this is an awkward result since such conception is being developed since 1950, by another side, it reveals the difficulty of the theme.

Key-Contribution. This paper's contribution is of theoretical and research interest. It shows that, despite the current expectation, whereas chatbots are being used to spread fake-news they are not being widely used to fight against them. This result is important because it is counter-intuitive. In addition, as this is unexpected, it suggests that may exist design difficulties hindering its use within this context. Such difficulties shall be identified and then overcome. This is a three-step procedure and this paper present the results for the first one.

Acknowledgments. This work has been supported by FCT – Fundação para a Ciência e Tecnologia within the Project Scope: UID/CEC/00319/2019.

References

1. Abdul-Kader, S.A., Woods, J.: Survey on chatbot design techniques in speech conversation systems. Int. J. Adv. Comput. Sci. Appl. **6**, 7 (2015)
2. Baldi, V., Gala, A.: A lógica crítica de c. s. peirce como processo de apoio na identificação de uma fake news. In: III International Conference Communicating Science in a Changing World (Universidade da Beira Interior - Covilhã, Portugal) (2018)
3. Chinnakotla, M.K., Agrawal, P.: Lessons from building a large-scale commercial IR-based chatbot for an emerging market. In: The 41st International ACM SIGIR Conference on Research & Development in Information Retrieval, USASIGIR 2018, pp. 1361–1362. ACM, New York (2018)
4. Daniel, F., Matera, M., Zaccaria, V., Dell'Orto, A.: Toward truly personal chatbots: on the development of custom conversational assistants. In Proceedings of the 1st International Workshop on Software Engineering for Cognitive Services, SE4COG 2018, pp. 31–36. ACM, New York (2018)
5. Dignum, F., Bex, F.: Creating dialogues using argumentation and social practices. Lecture Notes in Computer Science (including subseries Lecture Notes in Artificial Intelligence and Lecture Notes in Bioinformatics). LNCS, vol. 10750, pp. 223–235 (2018)
6. Ferrara, E., Varol, O., Davis, C., et al.: The rise of social bots. CACM **59**, 7 (2016)
7. Freitas, E., Bhintade, M.: Building Bots with Node.Js. Packt Pub, Birmingham (2017)
8. Friggeri, A., Adamic, L.A., Eckles, D., Cheng, J.: Rumor cascades. In: ICWSM (2014)
9. Hannak, A., Margolin, D., Keegan, B., Weber, I.: Get back! You don't know me like that: the social mediation of fact checking interventions in twitter conversations. In: ICWSM (2014)
10. Janarthanam, S.: Hands-On Chatbots and Conversational UI Development: Build Chatbots and Voice User Interfaces with Chatfuel, Dialogflow, Microsoft Bot Framework, Twilio, and Alexa Skills. Packt Publishing, Birmingham (2017)
11. Jurafsky, D., Martin, J., Norvig, P., Russell, S.: Speech and Language Processing. Pearson, London (2014)
12. Kitchenham, B., Brereton, P.: A systematic review of systematic review process research in software engineering. Inf. Softw. Technol. **55**(12), 2049–2075 (2013)
13. Klopfenstein, L.C., Delpriori, S., Malatini, S., Bogliolo, A.: The rise of bots: a survey of conversational interfaces, patterns, and paradigms. In: Proceedings of the 2017 Conference on Designing Interactive Systems, DIS 2017, pp. 555–565. ACM, New York (2017)
14. Kopp, C., Korb, K.B., Mills, B.I.: Information-theoretic models of deception: modelling cooperation and diffusion in populations exposed to "fake news". PloS ONE **13**, 11 (2018)
15. Kuhn, T.: The Structure of Scientific Revolutions. Foundations of the Unity of Science, vol. 2, no. 2. University of Chicago Press, Chicago (1962)
16. Li, T.J.-J., Riva, O.: Kite: building conversational bots from mobile apps. In: Proceedings of the 16th Annual International Conference on Mobile Systems, Applications, and Services, MobiSys 2018, pp. 96–109. ACM, New York (2018)
17. Marcondes, F.S., Almeida, J.J., Novais, P.: Chatbot Theory: A Naïve and Elementary Theory for Dialogue Management. LNCS. Springer, Heidelberg (2018)

18. Papaioannou, I., Lemon, O.: Combining chat and task-based multimodal dialogue for more engaging HRI: a scalable method using reinforcement learning. In: Proceedings of the Companion of the 2017 ACM/IEEE International Conference on Human-Robot Interaction, HRI 2017. ACM, New York (2017)

19. Parnas, D.: The real risks of artificial intelligence. Commun. ACM **60**, 10 (2017)

20. Pereira, J.: Leveraging chatbots to improve self-guided learning through conversational quizzes. In: Proceedings of the Fourth International Conference on Technological Ecosystems for Enhancing Multiculturality, TEEM 2016, pp. 911–918. ACM, New York (2016)

21. Peters, M., Rider, S., Hyvönen, M., Besley, T.: Post-Truth, Fake News: Viral Modernity & Higher Education. Springer, Singapore (2018)

22. Ramesh, K., Ravishankaran, S., Joshi, A., Chandrasekaran, K.: A survey of design techniques for conversational agents. In: International Conference on Information, Communication and Computing Technology. Springer, Heidelberg (2017)

23. Shao, C., Hui, P., Wang, L., Jiang, X., Flammini, A., Menczer, F., Ciampaglia, G.: Anatomy of an online misinformation network. PLoS ONE **13**, 4 (2018)

24. Stefanidi, E., Korozi, M., Leonidis, A., Antona, M.: Programming intelligent environments in natural language: an extensible interactive approach. In: Proceedings of the 11th PErvasive Technologies Related to Assistive Environments Conference, PETRA 2018, pp. 50–57. ACM, New York (2018)

25. Subramaniam, S., Aggarwal, P., Dasgupta, G.B., Paradkar, A.: Cobots - a cognitive multi-bot conversational framework for technical support. In: Proceedings of the 17th International Conference on Autonomous Agents and MultiAgent Systems, AAMAS 2018, pp. 597–604. International Foundation for Autonomous Agents and Multiagent Systems, Richland (2018)

26. Thakar, B., Parekh, C.: Advance persistent threat: botnet. In: Proceedings of the Second International Conference on Information and Communication Technology for Competitive Strategies, ICTCS 2016. ACM, New York (2016)

27. Vasconcelos, M., Candello, H., Pinhanez, C., dos Santos, T.: Bottester: testing conversational systems with simulated users. In: Proceedings of the XVI Brazilian Symposium on Human Factors in Computing Systems, IHC 2017, pp. 73:1–73:4. ACM, New York (2017)

28. Vo, N., Lee, K.: The rise of guardians: fact-checking URL recommendation to combat fake news. In: The 41st International ACM SIGIR Conference on Research, SIGIR 2018, vol. 38, pp. 275–284. ACM, New York (2018)

29. Wang, P., Angarita, R., Renna, I.: Is this the era of misinformation yet: combining social bots and fake news to deceive the masses. In: Companion Proceedings of the The Web Conference 2018, WWW 2018, pp. 1557–1561. International World Wide Web Conferences Steering Committee, Republic and Canton of Geneva (2018)

30. Yan, Z., Duan, N., Chen, P., Zhou, M., Zhou, J., Li, Z.: Building task-oriented dialogue systems for online shopping. In: Proceedings of the Thirty-First AAAI Conference on Artificial Intelligence, AAAI 2017. AAAI Press (2017)

31. Yuan, M.: Building Intelligent, Cross-platform, Messaging Bots, 1st edn. Addison-Wesley Professional, Boston (2018)

32. Zamanirad, S., Benatallah, B., Chai Barukh, M., Casati, F., Rodriguez, C.: Programming bots by synthesizing natural language expressions into API invocations. In: Proceedings of the 32nd IEEE/ACM International Conference on Automated Software Engineering, ASE 2017. IEEE Press (2017)

33. Zhou, X., Zafarani, R.: Fake news: a survey of research, detection methods, and opportunities. arXiv preprint arXiv:1812.00315 (2018)
34. Zubiaga, A., Aker, A., Bontcheva, K., Liakata, M., Procter, R.: Detection and resolution of rumours in social media: a survey. ACM Comput. Surv. (CSUR) **51**(2), 32 (2018)

Sale Forecast for Basic Commodities Based on Artificial Neural Networks Prediction

Jesús Silva[1(✉)], Jesús Vargas Villa[2], and Danelys Cabrera[2]

[1] Universidad Peruana de Ciencias Aplicadas, Lima, Peru
jesussilvaUPC@gmail.com
[2] Universidad de la Costa, St. 58 #66, Barranquilla, Atlántico, Colombia
{jvargas41,dcabrera4}@cuc.edu.co

Abstract. The objective of this paper is to carry out the comparison and selection of a method to forecast sales of basic food products efficiently. The source of data comes from a set of popular markets in the main departments of Colombia. The methods and methodologies used are: Hold Method, Winters, the Box Jenkins methodology (ARIMA) and an Artificial Neural Network. The results show that the artificial neural network obtained a better performance achieving the lowest mean square error.

Keywords: Sales forecast · Artificial Neural Networks (ANN) · Commodities

1 Introduction

Sales forecasts are indicators of economic-business realities, basically in the situation of industry in the market and in the participation of the company in that market. The forecast determines what can be sold with basis in reality, and the sales plan allows that hypothetical reality materializes, guiding the rest of the operational plans of the company. The choice and implementation of an appropriate method of forecasts has always been a big issue importance for companies. The forecasts in the area of purchases, marketing, sales, etc. A significant error in the forecast of sales could leave a company without the raw material or supplies necessary for its production or could generate an inventory Too big. In both cases, the prognosis wrong decreases the company's profits [1–3].

The optimal use of resources and the growing demand for a greater variety of products, among others, forces manufacturers to perform Stricter and flexible production schedules to be able to maximize the use of expensive production equipment, labor, investments in raw materials, so that the Delivery dates to final customer minimizing costs [4, 5].

Due to the nonlinear behavior of a sales forecast, artificial neural networks, ANNs, are an excellent candidate for the prediction of this estimate. ANNs are used in highly non-linear models and systems [6]. In general, ANNs are simple mathematical techniques designed to fulfill a wide variety of tasks. Nowadays the ANNs can be configured in various arrangements to develop diverse tasks, such as pattern recognition, data mining, classification and prediction, among others [7]. ANNs are composed of

© Springer Nature Switzerland AG 2020
F. Herrera et al. (Eds.): DCAI 2019, AISC 1003, pp. 37–43, 2020.
https://doi.org/10.1007/978-3-030-23887-2_5

attributes that learn solutions in applications where linear or non-linear mapping is needed. Some of these attributes are: ability to learn, generalization and parallel processing, these attributes make the ANNs can solve complex problems making this technique a precise and flexible method [8–10]. The objective of this article focuses on the use of neural networks to make sales forecasts and compare the results obtained against forecasts of classical statistical methods based on an average square error. Assuming that the use of a non-traditional forecasting method, such as neural networks, may provide a more accurate sales forecast compared to the results obtained using a traditional statistical forecasting method.

2 Theoretical Review

There are traditional models that are probabilistic, deterministic, and hybrids, such as: Simple Moving Average, Weighted Moving Average, Exponential Smoothing, Regression Analysis, Box-Jenkins method (ARIMA), trend projections, and other, used to generate forecasts providing certain advantages and disadvantages when compared to the other models. However, these traditional models are unable to offer good results in the current environment of high uncertainty and constant changes. For this reason, it is necessary to adopt new paradigms based on numeric modeling of nonlinear systems, like the Artificial Neural Networks (ANN), and the Support Vector Regression (SVR) [11].

The ANNs have shown a considerable development for reducing the error compared to traditional forecasting models. The difference lies in the fact that these new methods for analyzing indicators is based on the artificial learning, meaning that this algorithm acquires experience through the historical records that are entered as input values [12]. The demand for a product is generated by the interaction of several factors which are too complex to describe accurately with a mathematical model. So, this study proposes the use of the demand forecast.

Another method to achieve good results is the Support Vector Regression (SVR) which provides greater accuracy and improves the optimum global measure. Initially, the Support Vector Machines (SVM) were implemented for pattern recognition and classification [13], but currently, SVM has been implemented for linear regression (SVR). As examples of experiences with this method, [14] used this system to forecast sales of consumer products such as cars and perishable agricultural commodities respectively.

The use of more developed systems, such as ANN and Support Vector Machines have presented great improvements in forecasting as technology and more complex data management systems have evolved. An extensive review of scientific publications applying ANN on the price forecasting in the stock market around the world is exhibited by [15] this study concludes that the deep learning methods, compared to conventional models, show better results when increasing its accuracy considerably. However, they emphasize that these models present a difficulty in their structural conception because most cases are the result of a trial and error experimentation. In contrast, [16] confronted the resulting performance in time-series and diffuse logic with a structure based on time series, changing the inputs to the price variation and the trend sign, concluding that the stock market index of Taiwan provided more accurate results than the traditional forecasting models.

3 Materials and Methods

3.1 Data

In this study, the raw data were obtained from the National Administrative Department of Statistics of Colombia (DANE - Departamento Administrativo Nacional de Estadísticas), which provided a sales database of 1257 popular markets belonging to the SMEs sector in the main capitals of Colombia in the time period from 2015 to 2018 [17].

3.2 Description

The first stage to make forecasts is the collection of valid and reliable data. A forecast can not be more accurate than the data on which it is based. When a variable is measured over time, observations in different periods are often related or correlated. This correlation is measured by using the autocorrelation coefficient. Subsequently, the appropriate forecasting method is selected based on the pattern presented by the data, the type of series and the ease of application [18–20]. In addition, the appropriate parameters should be established for the type of forecast, this step was solved through exhaustive experimentation taking as a response variable a minimum mean square error [21].

4 Results

4.1 Autocorrelation Analysis

To avoid creating ambiguities when making a decision about the most convenient method when making forecasts, it is necessary to use a tool that allows evaluating the behavior of the series and that also considers the elements that compose it, i.e., that shows if the series is random, has a tendency or a seasonal pattern. A viable tool that was used for the autocorrelation analysis is the autocorrelogram [21, 22], with it is possible to establish an analysis of the pattern of the data series in addition to obtaining the autocorrelation coefficient, which will undoubtedly allow us to identify which elements of the series are present in this one Autocorrelation is the correlation that exists between a delayed variable with one or more periods.

Data patterns that include components such as trend, seasonality, and irregularity can be studied using the autocorrelation analysis approach. The autocorrelation coefficients for different time lags of a variable are used to identify patterns in the data time series. The following equation contains the formula for calculating the coefficient of autocorrelation (r_k) (see Ex. 1) between the observations X_t and X_{t-k}, the causal factors are at k distance periods [23].

$$r_k = \frac{\sum\limits_{t=1}^{T} (X_t - \bar{X})(X_{t-k} - \bar{X})}{\sum\limits_{t=1}^{T} (X_t - \bar{X})^2} \tag{1}$$

Where:

r = coefficient of autocorrelation for a delay of k periods.
= average of the values of the series. X_t = observations in period t.
X_{t-k} = observations of k previous periods or during a period t−k.

Applying Eq. 1, we obtain the coefficient of autocorrelation and in this way it is possible to verify the behavior of the time series Table 1.

Table 1. Autocorrelation function for each period

Delays of each period	Autocorrelation function
1	0.6542
2	0.4852
3	0.2458
4	0.2589
5	0.8246
6	−0.08763
7	−0.0428
8	0.04998
9	−0.0598
10	0.0652
11	−0.0658
12	0.05648

According to the results of Table 1, as the number of time delays increases, k, the autocorrelation coefficients decrease. In the results obtained, the data tend to fall gradually towards zero, in addition to the fact that the first delays are quite different from zero, as a result of the analysis of the data it is concluded that a trend pattern is present in the time series [24].

4.2 Selection of the Forecasting Method

The selection of the technique used to forecast, is defined based on the analysis of the factors that intervene in the selection of the model, in this case we look for suitable models that conform to trend patterns. According to [25], the suggested models to be used to carry out the forecasts are: Exponential smoothing adjusted to the trend, Holt's method and adjusted exponential smoothing for trend variations and seasonality, Winters Method in addition to the Box-Jenkins methodology (ARIMA), among others.

These statistical methods and methodologies will be compared against an artificial neural network which does not require testing an assumption of trend or seasonality in the time series prior to the realization of the forecast.

Table 2 shows the results obtained, including the mean absolute percentage error (MAPE). The data was divided into 8 clusters according to the type of products of first necessity [26–28]. The results were obtained from 800 repetitions of ANN training.

Table 2. Comparison of general results (percentages)

Methods	Cluster 1	Cluster 2	Cluster 3	Cluster 4	Cluster 5	Cluster 6	Cluster 7	Cluster 8	MAPE
Simple moving average (3)	10.85	11.84	12.54	9.54	12.5	14.8	13.5	11.14	12.58
Weighted moving average	42.5	43.52	44.25	44.58	42.58	45.5	44.5	44.69	44.12
Smoothing simple exponential	58.12	59.54	57.25	56.25	58.14	59.4	57.32	58.14	57.94
Linear regression	25.2	24.5	24.6	23.9	24.87	25.24	25.9	24.9	24.68
Seasonal or cyclical variation	52.2	54.2	56.2	54.8	54.9	51.2	57.2	54.2	54.84
Box-Jenkins - ARIMA (1,0,0)	14.5	14.2	14.2	16.5	14.5	14.98	14.5	12.98	14.91
ANN	9.58	8.25	8.64	8.34	8.14	8.95	8.26	8.24	**8.64**
(SVR) Support Vector Regression	12.5	12.4	12.6	12.8	12.7	12.5	12.9	12.4	12.35

The analyses carried out permit to observe that the best average value of MAPE in Table 2 is of ANN with 8.64%. As can be seen, the ANN model, compared to other more traditional forecasting models, present a considerably smaller error.

5 Conclusions

In the artificial neural network, it is not necessary to specify assumptions such as the probability distribution or behavior pattern in the time series for making predictions efficiently, since the method involves learning the relationships through the examples provided, which is the Contrary to conventional methods for forecasts where you have to specify the behavior pattern of historical data. According to the mean square error (MSE) as a performance metric, the artificial neural network presented a better practice when making predictions.

In the test phase carried out for three years, it can be seen that the total sales during that period of time are equivalent to: 44, 810, 911 lb. The predicted sales through the artificial neural network are: 44, 560, 866 lb. This reflects a 99.442% accuracy of the real sales vs forecast with the ANN, this propitiates a suitable scenario for the use of artificial neural networks when forecasting sales within the popular markets sector.

References

1. Amelec, V., Alexander, P.: Improvements in the automatic distribution process of finished product for pet food category in multinational company. Adv. Sci. Lett. **21**(5), 1419–1421 (2015)
2. Atsalakis, G., Valavanis, K.: Surveying stock market forecasting techniques – Part II: soft computing methods. Expert Syst. Appl. Part II **36**(3), 5932–5941 (2009)
3. Ayala, S.: La Economía como Ciencia, Objeto y Categorías Fundamentales (2015)
4. Garcia, M.: Análisis Y Predicción De La Serie De Tiempo Del Precio Externo Del Café Colombiano Utilizando Redes Neuronales Artificiales. Universitas Scientiarum **8**, 45–50 (2003)
5. Hanke, J., Wichern, D.: Pronósticos en los negocios. Pearson Educación, London (2006)
6. Matich, D.: Redes Neuronales: Conceptos básicos y aplicaciones. Cátedra de Informática Aplicada a la Ingeniería de Procesos–Orientación I (2001)
7. Lis-Gutiérrez, J.P., Lis-Gutiérrez, M., Gaitán-Angulo, M., Balaguera, M.I., Viloria, A., Santander-Abril, J.E.: Use of the industrial property system for new creations in Colombia: a departmental analysis (2000–2016). In: Tan, Y., Shi, Y., Tang, Q. (eds.) Data Mining and Big Data. DMBD 2018. Lecture Notes in Computer Science, vol. 10943. Springer, Cham (2018)
8. Obando, J.: Elementos de Microeconomía. EUNED (2000)
9. Ruan, D.: Fuzzy systems and soft computing in nuclear engineering. Physica (2013)
10. Sanclemente, J.: Las ventas y el mercadeo, actividades indisociables y de gran impacto social y económico.: El aporte de Tosdal. Innovar **17**(30), 160–162 (2007)
11. Sapankevych, N., Sankar, R.: Time series prediction using support vector machines: a survey. IEEE Comput. Intell. Mag. **4**(2), 24–38 (2009)
12. Toro, E., Mejia, D., Salazar, H.: Pronóstico de ventas usando redes neuronales. Scientia et technica **10**(26), 25–30 (2004)
13. Viloria, A., Robayo, P.V.: Inventory reduction in the supply chain of finished products for multinational companies. Indian J. Sci. Technol. **8**(1), 1–5 (2016)
14. Viloria, A., Gaitan-Angulo, M.: Statistical adjustment module advanced optimizer planner and SAP generated the case of a food production company. Indian J. Sci. Technol. **9**(47) (2016). https://doi.org/10.17485/ijst/2016/v9i47/107371
15. Villada, F., Muñoz, N., García, E.: Aplicación de las Redes Neuronales al Pronóstico de Precios en el Mercado de Valores. Información tecnológica **23**(4), 11–20 (2012)
16. Vitez, O.: Cuáles se consideran los principales indicadores económicos (2017). https://pyme.lavoztx.com/cules-se-consideran-los-principales-indicadores-econmicos-9641.html
17. Wu, Q., Yan, H., Yang, H.: A forecasting model based support vector machine and particle swarm optimization. In: 2008 Workshop on Power Electronics and Intelligent Transportation System, pp. 218–222 (2008)
18. Zhang, G.: Time series forecasting using a hybrid ARIMA and neural network model. Neurocomputing **50**(Suppl. C), 159–175 (2003)
19. Deliana, Y., Rum, I.A.: Understanding consumer loyalty using neural network. Pol. J. Manag. Stud. **16**(2), 51–61 (2017)
20. Scherer, M.: Waste flows management by their prediction in a production company. J. Appl. Math. Comput. Mech. **16**(2), 135–144 (2017)
21. Chang, O., Constante, P., Gordon, A., Singana, M.: A novel deep neural network that uses space-time features for tracking and recognizing a moving object. J. Artif. Intell. Soft Comput. Res. **7**(2), 125–136 (2017)

22. Sekmen, F., Kurkcu, M.: An early warning system for Turkey: the forecasting of economic crisis by using the artificial neural networks. Asian Econ. Financ. Rev. **4**(1), 529–543 (2014)
23. Ke, Y., Hagiwara, M.: An English neural network that learns texts, finds hidden knowledge, and answers questions. J. Artif. Intell. Soft Comput. Res. **7**(4), 229–242 (2017)
24. Cai, Q., Zhang, D., Wu, B., Leung, S.C.: A novel stock forecasting model based on fuzzy time series and genetic algorithm. Procedia Comput. Sci. **18**(1), 1155–1162 (2013)
25. Egrioglu, E., Aladag, C.H., Yolcu, U.: Fuzzy time series forecasting with a novel hybrid approach combining fuzzy c-means and neural networks. Expert Syst. Appl. **40**(1), 854–857 (2013)
26. Kourentzes, N., Barrow, D.K., Crone, S.F.: Neural network ensemble operators for time series forecasting. Expert Syst. Appl. **41**(1), 4235–4244 (2014)
27. Departamento Administrativo Nacional de Estadística-DANE.: Manual Técnico del Censo General. DANE, Bogotá (2018)
28. Fajardo-Toro, C.H., Mula, J., Poler, R.: Adaptive and hybrid forecasting models—a review. In: Ortiz, Á., Romano, A.C., Poler, R., García-Sabater, J.P. (eds.) Engineering Digital Transformation. Lecture Notes in Management and Industrial Engineering. Springer, Cham (2019)

Predicting Shuttle Arrival Time in Istanbul

Selami Çoban[1]([✉]), Victor Sanchez-Anguix[3], and Reyhan Aydoğan[1,2]

[1] Department of Computer Science, Özyeğin University, Istanbul, Turkey
selami.coban@ozu.edu.tr, reyhan.aydogan@ozyegin.edu.tr
[2] Interactive Intelligence Group, Delft University of Technology,
Delft, The Netherlands
[3] Florida Universitaria, Carrer del Rei en Jaume I, Catarroja,
Valencia, Spain
vsanchez@florida-uni.es

Abstract. Nowadays, transportation companies look for smart solutions in order to improve quality of their services. Accordingly, an intercity bus company in Istanbul aims to improve their shuttle schedules. This paper proposes revising scheduling of the shuttles based on their estimated travel time in the given timeline. Since travel time varies depending on the date of travel, weather, distance, we present a prediction model using both travel history and additional information such as distance, holiday, and weather. The results showed that Random Forest algorithm outperformed other methods and adding additional features increased its accuracy rate.

Keywords: Smart cities · Transportation · Information Fusion · Data Science

1 Introduction

Despite having some advantages, living in large metropolitan cities also has some disadvantages. Due to the enormous number of vehicles, people spend significant amount of their daily time in traffic. Since punctuality is one of the vital components for transportation services, transportation companies need to find methods for reducing waiting time or properly informing customers of expected delays [1], so that they can plan ahead. In the end, transportation companies need to establish good relationships with their passengers. This paper particularly focuses on how bus companies can employ the predictive power of machine learning to improve the quality of shuttle services provided by intercity buses. More specifically, we focus on a real case scenario based on the city of Istanbul.

In Istanbul, intercity bus companies usually have a central station or hub, which is located out of the city center. Since there may not be direct public transportation to these hubs, companies provide shuttle services to pick up their passengers from certain populated areas. Scheduling of these shuttles is

© Springer Nature Switzerland AG 2020
F. Herrera et al. (Eds.): DCAI 2019, AISC 1003, pp. 44–51, 2020.
https://doi.org/10.1007/978-3-030-23887-2_6

important since any latency will cause a delay in the intercity bus schedule. Furthermore, if the shuttles arrive at the station too early, passengers have to wait for long time, which would result in decreasing customer satisfaction. The schedules are usually set and fixed by a human operator by considering the distance between the picking area and the station as well as the average/worst-case travel time in Istanbul.

Without a doubt, there are additional factors influencing travel time such as time of day (i.e., whether or not it is a rush hour), date of travel (i.e., workday, weekends, school days, public holidays), weather (i.e., sunny, rainy, snowy, fogy etc), and so on. Therefore, planning without taking those factors into account would not yield a satisfactory result for passengers. For this problem we aim to design an analytics system that uses machine learning methods to predict travel time of shuttles by using history of previous trips, as well as taking the aforementioned factors into consideration. The aim of the framework is to revisit scheduling suitably.

This paper investigates what information should be used by learning algorithms to give more accurate estimations. Three different machine learning algorithms, Random Forests, Support Vector Machines, and Deep Neural Networks have been tested on the real data provided by an intercity bus company. The results show that, for this first batch of experiments, Random Forests outperform other methods. On top of that, we also identified that weather and holiday information have a considerable impact on the accuracy of the predictions.

The rest of the paper is organized as: Sect. 2 explains the proposed framework elaborately while Sect. 5 presents our experimental design and empirical evaluations of our findings. A list of related work is given in Sect. 6. Lastly, we summarize our contributions and discuss future work in Sect. 7.

2 Proposed Approach

In order to increase passenger satisfaction, intercity bus companies provide shuttle services to hubs where intercity buses depart/arrive in Istanbul. Human operators usually schedule shuttles in advance based on their knowledge. This process is more complicated than it seems. On the one hand, any latency in arrivals of shuttles to the hub will cause delays in intercity bus schedules. On the other hand, if passengers are picked up too early, then they have to wait for the bus in the hub. Waiting for long time may frustrate passengers, which decreases customer satisfaction too. Optimally scheduling shuttles requires knowing how long the trip will last in advance. However, the travel duration in metropolitan cities such as Istanbul varies depending on multiple factors such as the weather condition, working hours, school-periods, and so forth. In Sect. 3, we propose the architecture for an analytics system whose aim is that of dynamically scheduling shuttles provided by intercity bus companies. In latter sections, we present the design and experiments for the predictive modules of the system.

3   Analytics System Architecture

Instead of using some predefined time constraints in the schedule algorithm, we propose to schedule shuttles based on estimated travel duration by a machine learning algorithm as depicted in Fig. 1. The system consists of the following modules:

- Historical data source: This constitutes the main source of information for the analytics system. It contains the past and present data about the trips carried out by the shuttles. It is an internal source of information for the transportation company.
- External data source: It contains information from external sources that is publicly available. For instance, data about the weather that can be obtained for free from API REST services.
- Feature transformation: This module is in charge of transforming data collected from internal and external sources into an appropriate format for machine learning algorithms. Travel information might be recorded by the drivers or through some sensors. Despite this, drivers may forget to record their arrival time on time; consequently, duration of the trip might seem unexpectedly long. Similarly, any technical problem in the sensors may cause noisy data as well. Before training the model, we need to clean noisy data. In addition to this, new features can be engineered by processing the existing data.
- Information fusion: This module links the information from the multiple sources into a single dataset that can be employed by the machine learning algorithms.
- Prediction engine: In this module, a request is provided in the form of a future trip and historical data. The machine learning algorithm will produce an estimation of the expected arrival time to the central hub.
- Scheduling module: The scheduling module employs scheduling algorithms to determine how shuttles should be organized and timetabled throughout the year. For that, it uses predictions provided by the Prediction engine.

4   Preparing the Prediction Engine

In this section we describe the steps that we carried out in order to train the predictive model employed in the prediction engine. For that, we use an internal dataset consisting of shuttle trip information for 132 days recorded by an intercity bus company. Out data set consists of 32888 records including the following information: *hub location, route, type of trip, scheduled departure time, departure time, arrival time, vehicle id* and *number of stop*. There are two types of trip: picking up from local areas to the central station and delivering to those areas from the central station. There are 11 different routes (i.e. names of the main local areas). It is worth noting that shuttles have to make some stops prior to arriving to the hub. Next, we describe the general methodology used for training the machine learning models.

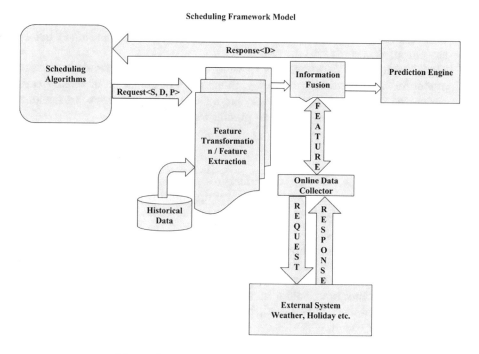

Fig. 1. Proposed shuttle scheduling framework

- Cleaning noisy data: In the given data set, there were some outliers. We sampled the existing data set in order to remove potential outliers.
- Feature selection: Relevant features for predicting the trip duration are determined and irrelevant features are eliminated from the training data set. For example, we eliminate "central station" since there is only one central station, which is the same for all data instances. Some features are preprocessed and transformed into new features capturing more meaningful information. For instance, one hot coding is used for arrival time. Since our regression model aims to predict trip duration, an output feature of trip duration in seconds is added to the training data set by calculating the difference between departure time and arrival time.
- Enriching the dataset via Information Fusion: The feature set has been extended with the additional information such as distance (i.e., how far a center from arrival point), weather and holiday information gained from public services. Our final data set includes the following features: route (categorical data), type of route (categorical data), number of stop, distance, hour (integer value between 0 and 23), day of the week (integer value between 0 and 6 where 0 denotes Monday while 6 denotes Friday), rain in terms of millimeter (amount of rain on the given date), snow in terms of centimeter (amount of snow in the given date), is holiday (1 denotes if the given date is public holiday in Turkey; otherwise it is 0), and is weekend (1 denotes if the

given date is weekend; otherwise it is 0). Note that the amount of rain and snow crawled using script in Accuweather Web site[1]. As mentioned before, the output feature is trip duration.
- Model training: It is important to choose the most appropriate regression model for high prediction accuracy. Since our data consists of mostly categorical data and we do not have very large data set, we propose to use random forest algorithms for regression to predict travel time [2].

5   Evaluation

In order to study the effect of which features play key role in the performance of the given regression algorithm, eight training data sets are created by using different feature combinations and encoding as shown below.

1. # of stop and distance.
2. # of stop, distance, and hour.
3. # of stop,distance,is weekend, and hour.
4. # of stop,distance,is weekend, hour, and day of week.
5. # of stop,distance,is weekend, hour, day of week, rain, snow, and is holiday.
6. # of stop, distance, is weekend, hour, day of week, rain, snow, is holiday, route and route type.
7. Feature set 6 but one hot encoding is used for hour.
8. Feature set 6 but one hot encoding is used for hour and day of week.

We test the performance of the regression algorithm for each data set to see whether using additional features increase the accuracy of the predictions. Furthermore, we compare the performance of the Random Forest Algorithm with SVM [3] and Deep Learning [4] algorithm for each feature set. SVM and RF Models are trained by 10 fold cross validation technique. Deep Learning algorithm has been trained by splitting data samples 20% validation set. We carried out a grid search on the space of hyperparameters for each algorithm. Due to space constraints, we only report the results of the best hyperparameter configurations.

In case of the best results found for the Random Forest implementation, bootstrapped samples were 200, the max depth was 6, and minimum split parameter was set to 4. In the SVM implementation, RBF kernel was employed. With regards to the deep learning approach, a multi-layer deep network containing two hidden layer was found to be the best. The first hidden layer in network had 512 units while the second hidden layer contained 256 units. The activation function in each layer was RELU function. The model has been trained via 512 samples batch mode for 150 epoch in each feature sets. The best learning rate was found at 0.001.

In our experiments, we evaluated the prediction performance of each learning algorithm for each feature set according to a variety of evaluation metrics widely used in the literature. The following performance metrics were calculated in our experiments:

---

[1] https://www.accuweather.com/tr/tr/istanbul/318251/.

– **Mean Squared Error:** The Mean square error is a measure of the quality of an estimator. It is always non-negative, and values closer to zero are better.

$$\frac{1}{n}\sum_{t=1}^{n}(Y_{real}^{t} - Y_{prediction}^{t})^{2} \tag{1}$$

– **Mean Absolute Error:** The Mean absolute error is a method to estimate the performance of model. The method simply calculates the absolute error between actual and prediction. It's not that sensitive to outlier as mean square error.

$$\frac{1}{n}\sum_{t=1}^{n}|(Y_{real}^{t} - Y_{prediction}^{t})| \tag{2}$$

– **Mean Squared Log Error:** The error metric simply checks the difference between actual and prediction. For example for P = 500 and A = 250 would give you the roughly same error as when P = 5000 and A = 2500.

$$\frac{1}{n}\sum_{t=1}^{n}(\log(Y_{prediction}^{t} + 1) - \log(Y_{real}^{t} + 1))^{2} \tag{3}$$

– **Median Absolute Error:** The MAE is particularly interesting because it is robust to outliers. The loss is calculated by taking the median of all absolute differences between the target and the prediction.

$$median(|(Y_{prediction} - Y_{real})|) \tag{4}$$

– **Mean Absolute Percentage Error:** MAPE measures the size of the error in percentage terms. It is calculated as the average of the percentage error.

$$(\frac{1}{n}\sum_{t=1}^{n}|\frac{Y_{real}^{t} - Y_{prediction}^{t}}{Y_{real}^{t}}|) * 100 \tag{5}$$

Figure 2 shows the average errors of Random Forest, SVM and Deep learning algorithm on each feature set. The result shows that Random Forest learning model outperformed both Support Vector Machine and Deep Learning, and SVM performed better than Deep learning. The low performance of deep learning may stem from working with a small-sized data set.

It can be obviously observed that Random Forest gained the best performance on the feature set 6 with respect to all metrics. Surprisingly, the best performance is gained with the feature set 1 where no additional information is used. In deep learning case, except MAPE metric (feature set 4), feature set 8 reveals the best performance where one hot encoding was used for hour and day of week. In conclusion, we observed increase in prediction accuracy when we add weather, distance and holiday information.

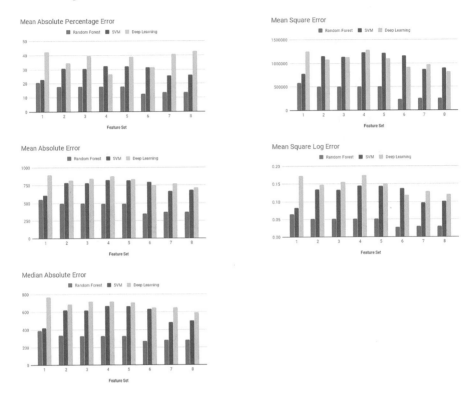

Fig. 2. Average errors for each algorithm on each feature set

6   Related Work

Zhang and Haghani studied travel time prediction in Maryland [5]. They compare the performance of Gradient Boosting Regression Tree method (GBM) with other methods such as auto-regressive integrated moving average and Random Forest (RF). The results showed that both RF and GBM are promising for travel time predictions. Furthermore, Kumar *et. al* focus on real time travel prediction and use both temporal and spatial variations and show that it performs better than historical average, regression, and ANN methods and only one of those temporal or spatial variations [6].

Schnitzler *et. al* propose a prediction model using both historical journeys and real time data streamed from bus in city of Dublin [7]. A combination of several methods from Queuing Theory and Machine Learning in the prediction has been studied. Another study builds a prediction model by using GPS data and artificial neural network (ANN) [8]. That study considers number of passengers, average nonstop trip duration, number of dwells at each stop. The performance of ANN was compared with Linear Regression Model and ANN slightly performed better. Kuchipudi and Chien develop an adaptive algorithm using two models: linked based and path based models in order to improve the accuracy of travel time prediction [9].

7 Conclusion and Future Works

This paper introduces a framework for effective scheduling of shuttles by inter city bus companies. Particularly, the focus is to predict travel time of shuttles in an accurate way so that we can make a better schedule. Moreover, we aim to increase the prediction accuracy by enriching the feature set by adding additional information such as weather, distance, and holiday. Our experimental results showed that adding those features increases the prediction accuracy. In addition, we compared three regression algorithm in this setup. It is observed that Random Forest outperformed SVM and deep learning algorithm. As a future work, we plan to design an effective schedule algorithm using predicted travel time.

References

1. Aydoğan, R., Lo, J.C., Meijer, S.A., Jonker, C.M.: Modeling network controller decisions based upon situation awareness through agent-based negotiation. In: Frontiers in Gaming Simulation, pp. 191–200. Springer, Heidelberg (2014)
2. Louppe, G.: Understanding random forest. Ph.D. dissertation, University of Liège (2015)
3. Alpaydin, E.: Introduction to Machine Learning, 2nd edn, pp. 309–335 (2010)
4. Goodfellow, Y.B., Courville, A.: Deep learning (2017)
5. Zhang, Y.Y., Haghani, A.: A gradient boosting method to improve travel time prediction. Transp. Res. Part C: Emerg. Technol. **58**, 308–324 (2015)
6. Kumar, B., Vanajakshi, L., Subramanian, S.: Bus travel time prediction using a time-space discretization approach. Transp. Res. Part C: Emerg. Technol. **79**, 308–332 (2017)
7. Schnitzler, F., Senderovich, A., Gal, A., Mandelbaum, A., Weidlich, M.: Traveling time prediction in scheduled transportation with journey segments. Inf. Syst. **64**, 266–280 (2016)
8. Amita, J., Jain, S., Garg, P.: Prediction of bus travel time using ann: a case study in delhi. Transp. Res. Part C: Emerg. Technol. **17**, 263–272 (2016)
9. Kuchipudi, C., Chien, S.: Development of a hybrid model for dynamic travel-time prediction. Transp. Res. Rec.: J. Transp. Res. Board **1855**, 22–31 (2003)

Human-Computer Interaction in Intelligent Tutoring Systems

Ramón Toala[1,2] ⓘ, Dalila Durães[1,3,4(✉)] ⓘ, and Paulo Novais[1] ⓘ

[1] Algoritmi Research Centre/Department of Informatics, University of Minho,
Braga, Portugal
id7410@alunos.uminho.pt, pjon@di.uminho.pt
[2] Technical University of Manabí, Portoviejo, Manabí, Ecuador
[3] CIICESI, ESTG, Polytechnic Institute of Porto, Felgueiras, Portugal
dad@estg.ipp.pt
[4] Department of Artificial Intelligence, Technical University of Madrid,
Madrid, Spain

Abstract. Due to the rapid evolution of society, citizens are constantly being pressured to obtain new skills through training. The need for qualified people has grown exponentially, which means that the resources for education/training are significantly more limited, so it's necessary to create systems that can solved this problem. The implementation of Intelligent Tutoring Systems (ITS) can be one solution. Besides, ITS aims to enable users to acquire knowledge and develop skills in a specific field. To achieve this goal, the ITS should learn how to react to the actions and needs of the users, and this should be achieved in a non-intrusive and transparent way. In order to provide personalized and adapted system, it is necessary to know the preferences and habits of users. Thus, the ability to learn patterns of behaviour becomes an essential aspect for the successful implementation of an ITS. In this article, we present the student model of an ITS, in order to monitor the user's biometric behaviour and their learning style during e-learning activities. In addition, a machine learning categorization model is presented that oversees student activity during the session. Additionally, this article highlights the main biometric behavioural variations for each activity, making these attributes enable the development of machine learning classifiers to predict users' learning preferences. These results can be instrumental in improving ITS systems in e-learning environments and predict user behaviour based on their interaction with computers or other devices.

Keywords: Intelligent Tutoring Systems · Human-computer interaction · Behaviour biometrics

1 Introduction

ITS are learning environments that help students in the teaching-learning process. ITS implements intelligent algorithms that adapt to users and allow the application of complex learning principles. An ITS should normally work with only one user because

© Springer Nature Switzerland AG 2020
F. Herrera et al. (Eds.): DCAI 2019, AISC 1003, pp. 52–59, 2020.
https://doi.org/10.1007/978-3-030-23887-2_7

users differ in many dimensions and the goal is to be sensitive to the idiosyncrasies of individual users. Some basic ITS activities should incorporate active user learning, interactivity, adaptability, and feedback.

However, we know that ITS's are complex computer programs that manage several heterogeneous types of knowledge. Thus, building such a system is therefore not an easy task. It is necessary that the team that will construct the ITS be multidisciplinary to face several problems related to the construction process. In fact, the necessary resources to build an ITS comes from various disciplines, including artificial intelligence, cognitive science, science education, human-computer interaction and software engineering. This multidisciplinary approach makes the process of building an ITS a challenging task since each researcher can have very different views of the system. Some promote the pedagogical accuracy, while others focus on effective diagnosis of the errors of the students [1].

Considering the points described so far, the objective of this paper, is to present the student model of an ITS in order to monitor the user's biometric behaviour and their learning style during e-learning activities. In addition, a machine learning categorization model is presented that oversees student activity during the session. This paper is organized as follows: section two contains the theoretical foundations; section three has the proposed architecture; section four presented the methodology applied and some results; finally, in section five the conclusions of this work are presented and some future work.

2 Theoretical Foundations

2.1 ITS

The ITS arose from the authors of the research with experience in three different disciplines: computer science (AI), psychology (cognition) and education (education and training) [2]. An ITS is a platform that incorporates AI techniques to create tutors. An ITS is an "automatic" platform, which draws courses for a person and adjusts content according to circumstances [3].

Typically, the architecture of an ITS have four essential components: the expert model, the student model, the tutor model and the interface. The expert model contains the rules for the representation of concepts, rules and problem-solving strategies. Additionally, this model highlights the cognitive and affective states of the user in association with their progression as the learning process proceeds. The tutor model is the part of the ITS that designs and regulates instructional interactions with the user. This model accepts information from the student model and the expert model [3]. The student model is an overlap of the expert model. The interface is the front-end interaction with the user. So, the ITS integrates all types of information needed to interact with the user through graphics, text, multimedia, video, menus, and others [3].

The ITS consists of a system that applies the techniques of Ambient Intelligence (AmI) to better provide and support users of the educational system [4]. However, what is observed in most of these systems is that they are either adaptive or intelligent, but not adaptive and intelligent [5].

2.2   Affective Computing

The concept of Affective Computing (AC) was introduced by Picard [6], who defined it as: "Computation that relates to, arises or deliberately influences emotions." The AC focuses on the establishment of models based on physiological and behavioural signals collected by sensors and techniques to understand, recognize, and understand human emotions and provide better feedback [7].

There are several fields in which AC can be applied: HCI; text-based communication and virtual reality; human monitoring; and education systems. In the field of HCI, the machine needs to feel and react to human emotion respectively. For example, if a person communicates with technology and is feeling frustrated or confused, technology needs to be able to respond differently to that person depending on their emotional state [8]. Thus, it is possible to build a personalized computer system with the ability to perceive, interpret the feelings of the human being, as well as giving us intelligent, sensitive and adapted responses to situations [6].

One of the problems in studying affects is the definition of what is emotion. Some research was intended to relate emotion and computer, so for Ortony, Clore and Collins [9] the identification of emotions is generally used in the field of cognitive science and has a connection to affective computing allowing computers to recognize and express emotions [6].

2.3   Pattern Interaction Using Behaviour Biometrics

A method for obtaining a high set of data is through the use of the interactions of the keyboard and mouse. These interactions can provide a non-intrusive and easy-to-use continuous monitoring method. The behaviour of the mouse and the way we click the mouse buttons and the movement we make with the mouse changes in response to various factors, such as stress and health status [10, 11]. Based on the keystroke typing patterns and the mouse movement pattern, the researchers also observed a certain inherent variation in the typing pattern of an individual [12].

Mouse and keyboard tracking are techniques that are also used to measure and rank attention. These techniques have already been used to measure other variables, such as stress [12] and mental fatigue [13]. These techniques are a non-invasive approach because the captured data is made by software running in the background and the user does not have the perception that it is being monitored. This is an advantage over other approaches because users have no idea that they are being supervised and do not change their default behaviour.

2.4   Attention and Learning Styles

Attention means concentration on one of the various subjects or objects that can capture our minds to the detriment of others [15]. Thus, the concept of attention can also be defined as the change of a huge set of unstructured data acquired in structured data, where the main information is preserved. Previous results indicate that learning styles can influence learning performance [16]. That is, each user has his own way of assimilating knowledge, his/her preference of learning and when this style is known, the learning process is maximized.

3 Proposed Architecture

Based on the information provided in section two, the idea is to create an ITS system tailored to each user. The student model is a subsystem that discovers user's common behaviours and habits from data recorded by sensors. The main of the system is the learning algorithm, which combined with a language, allows to discover and to represent the patterns. This algorithm is divided in four steps which represent the four logical steps to discover frequent and comprehensible patterns presented in Fig. 1.

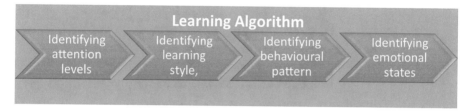

Fig. 1. Algorithm of student model of an ITS.

Based on the algorithm of Fig. 1, we present the proposed architecture of the ITS student model. To develop this model, it was necessary to define the following parameters: attention levels, learning style, user behaviour style and emotional state. After these defined parameters, the student model of the ITS should adjust and adapt the level of learning difficulty to each user, depending on their parameters. Thus, it was necessary to develop a new architecture due to the void found in the literature review, since most of the systems developed to date are invasive and intrusive. The idea of creating a non-intrusive and non-invasive approach based on the observation of behavioural changes of an individual or a group in relation to the behavioural and emotional pattern emerged.

The structure of the proposed architecture is shown in Fig. 2. This proposed architecture is based on the traditional ITS, with the four main modules: the expert model, the tutor model, the student model, and the interface. The student model is subdivided into two submodules: the student style and the state of knowledge. The submodule student style is where all information about the level of attention, learning style and emotional state are classified and stored. The submodule state of knowledge is received all the information from the interface, update the state updater and based on the current state of knowledge, the learning style, the level of attention and the emotional state, provide a recommendation system in order to adapt for the user in order to improve the learning process [16].

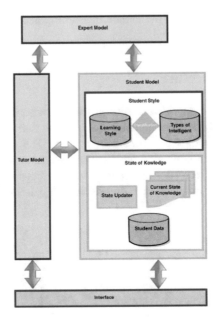

Fig. 2. Main modules of an ITS [16].

4 Methodology and Results

In this work, the approach followed is based on the dynamics of the mouse and the keys, in an attempt to propose a completely non-intrusive method to evaluate the human-computer interaction. For this study, professional students were selected to participate as they are students who easily give up activities and therefore require further monitoring.

4.1 Population

The study was carried out at High School of Caldas das Taipas, where evaluation activities are carried out in the computer with no intervention of the teacher. Each computer has a keyboard, a mouse, and a screen. The assessment activity starts at the same time for all students and they log in to the default software using their personal credentials and the activity begins. The experiment took place in four different lessons, each lasting 100 min. In each lesson, different learning styles were applied to the same content in order to determine the user learning style.

The first lesson used video exercises with step-by-step instructions. The second lesson used exercises with pictures, where the students would have to arrive at this information. The third lesson, used exercises with only text, without any supporting image. Finally, the fourth lesson used exercises with indications only audio. For this purpose, a group of 14 students were selected to carry out this experiment. Information on the four lessons is presented in Table 1.

Table 1. Summary of the characteristics of each assess activity.

Class	Date	Duration (minutes)		
		x	$\tilde{x}$	S
Video	21-05-2018	7.31612E + 15	8.4653E + 15	3.00389E + 15
Image	24-05-2018	9.99167E + 15	1.21798E + 16	4.7135E + 15
Text	25-05-2018	6.19021E + 15	7.29301E + 15	2.64201E + 15
Audio	29-05-2018	7.0459E + 15	8.95808E + 15	3.55391E + 15

5 Dataset

After completing the four lessons with different learning styles, a small questionnaire was applied to students to indicate their preferred learning style. The result indicated: for 70% for the image and 30% for the video. Considering these results, considering the results obtained in the evaluation of the exercises performed in the four lessons, in the data presented in Table 2 and in Fig. 3 we can conclude that: (a) the best level of

Table 2. Summary of the biometric feature's variation for each lesson.

Class	CD	DBC	DDC	DPLBC KDT	MA	MV	APP	TBK
Units	ms	px	px	ms px	px/s2	px/ms	%	ms
Video	188.31	206.1	214.24	2819.12 1002.8	0.6811	0.5931	94.14	1865
Image	150.73	150.73	161.88	3725.35 2669.1	0.6659	0.5413	91.36	1738
Text	146.43	227.4	152.52	2838.14 1223.0	0.6056	0.5312	89.23	1073.1
Audio	281.0	189.2	119.10	2408.52 714.1	0.5828	0.4986	93.7	1758
$\bar{x}$	187.66	227.0	165.70	2946.41 1384.2	0.6367	0.5447	92.08	1599.3

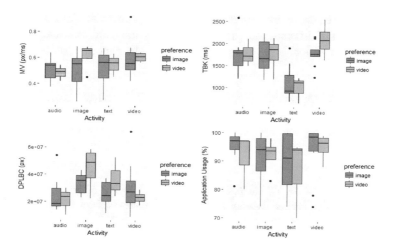

Fig. 3. Activity for the four lessons.

attention was obtained in the video lesson, which was the favorite learning style of only 30% of the students. In the students' favorite style, students had a level of attention of 91.36, which has a lowest average. (b) In the case of mouse speed, the highest level was obtained in the video and image lesson.

6   Conclusion and Future Work

This paper only presents the student model of an ITS system. A non-invasive and non-intrusive approach to an ITS is proposed based on behaviour biometric analysis of work in different classes with some different learning styles. The system monitors and analyzes the dynamics of the mouse, the dynamic of the keyboard and tasks to determine user performance.

The results show that it is still possible to improve some mechanisms to better understand the relationship between human behaviour, attention and evaluation in order to implement better learning strategies for each user. These results are crucial to improving learning systems in an e-learning environment and anticipate user behaviour based on their interaction with technology devices.

Succinctly, it was verified that the key characteristics to be analyzed to obtain a prediction about the user, concentrate on the Key Down Time, Mouse Velocity, and Distance Point between Clicks (DPLBC). However, since the favorites indicated by the user were only of the "video" and "image" activities, it is necessary to make new case studies to better complement the other activities ("audio" and "text"). Increasing the number of cases to be applied in the learning process, the greater the validation of the future model.

As future work, still in the development of the student model, the research will be focused on: (a) increasing the number of case studies available for analysis; (b) increase the number of quality resources that would allow better monitoring of the user's attention performance; (c) detailed analysis of characteristics that influence user's performance (for example, by correlating user' final grades with biometric behaviors); and (d) definition of different user profiles to improve the platform's adaptive learning mechanisms. In addition, it is necessary to develop other models to obtain complete ITS.

Acknowledgement. This work has been supported by FCT – Fundação para a Ciência e Tecnologia within the Project Scope: UID/CEC/00319/2019.

References

1. O'Donnell, E., Lawless, S., Sharp, M., Wade, V.P.: A review of personalised e-learning: towards supporting learner diversity. Int. J. Distance Educ. Technol. (IJDET) **13**(1), 22–47 (2015)
2. Cataldi, Z., Lage, F.J.: Sistemas tutores inteligentes orientados a la enseñanza para la comprensión. Edutec. Revista Electrónica de Tecnología Educativa **0**(28), 108 (2009). https://doi.org/10.21556/edutec.2009.28.456

3. Ahuja, N.J., Sille, R.: A critical review of development of intelligent tutoring systems: retrospect, present and prospect. Int. J. Comput. Sci. Issues (IJCSI) **10**(4), 39 (2013)
4. Brusilovsky, P., Peylo, C.: Adaptive and intelligent webbased educational systems. Int. J. Artif. Intell. Educ. **13**(2–4), 159–172 (2003). http://dl.acm.org/citation.cfm?id=1434845. 1434847. ISSN 1560-4292
5. Rodrigues, M., Novais, P., Santos, M.: Future challenges in intelligent tutoring systems – famework, recent research developments in learning technologies. In: Méndez Villas, A., Gonzalez Pereira, B., Mesa González, J., Mesa González, J.A. (eds.) Proceedings of the 3rd International Conference on multimedia and Information & Communication Technologies in Education, pp. 929–934. Publishers Formatex (2005)
6. Picard, R., Papert, S., Bender, W., Blumberg, B.: Affective learning - a manifesto. BT Technol. J. **22**(4), 253–268 (2004)
7. Lee, H., Choi, Y., Lee, S., Park, I.: Towards unobtrusive emotion recognition for affective social communication. In: The 9th Annual IEEE Consumer Communications and Networking Conference - Special Session Affective Computing for Future Consumer Electronics. IEEE, Las Vegas (2012)
8. Pang, B., Lee, L.: Opinion mining and sentiment analysis. Found. Trends® Inf. Retrieval **1**(2), 91–231 (2006)
9. Ortony, A., Clore, G., Collins, A.: The Cognitive Structure of Emotions. Cambridge Press, Cambridge (1990)
10. Monrose, F., Rubin, A.: Keystroke dynamics as a biometric for authentication. Future Gener. Comput. Syst. **16**(4), 351–359 (2000)
11. Araújo, L.C., Sucupira, L.H., Lizarraga, M.G., Ling, L.L., Yabu-Uti, J.B.: User authentication through typing biometrics features. IEEE Trans. Sig. Process. **53**(2), 851–855 (2005)
12. Carneiro, D., Novais, P., Pêgo, J., Sousa, N., Neves, J.: Using mouse dynamics to assess during online exams. Hybrid Artif. Intell. Syst. **9121**, 345–356 (2015)
13. Pimenta, A., Gonçalves, S., Carneiro, D., Riverola, F., Novais, P.: Mental workload management as a tool in e-learning scenarios. In: B-Peces, Paillet, O., Ahrens, A. (eds.) Proceedings of the 5th International Conference on Pervasive and Embedded Computing and Communication Systems, pp. 25–32. Scite Press (2015)
14. Mancas, M.: Attention in computer science - part 1. News and insights from EAI community – Blog, 6 October 2015. http://blog.eai.eu/attention-in-computer-science-part-1/. Accessed 31 Dec 2016
15. Toala, R., Gonçalves, F., Durães, D., Novais, P.: Adaptive and intelligent mentoring to increase user attentiveness in learning activities. In: Simari, G., Fermé, E., Gutiérrez Segura, F., Rodríguez Melquiades, J. (eds.) Advances in Artificial Intelligence - IBERAMIA 2018. IBERAMIA 2018. Lecture Notes in Computer Science, vol. 11238. Springer, Cham (2018)
16. Andrade, F., Novais, P., Carneiro, D., Zeleznikow, J., Neves, J.: Using BATNAs and WATNAs in online dispute resolution. In: Nakakoji, K., Murakami, Y., McCready, E. (eds.) New Frontiers in Artificial Intelligence, JSAI-isAI 2009. Lecture Notes in Computer Science, vol. 6284. Springer (2010). http://dx.doi.org/10.1007/978-3-642-14888-0_2

Bioinformatics, Biomedical Systems, e-health, Ambient Assisting Living

Smart Cities: A Taxonomy for the Efficient Management of Lighting in Unpredicted Environments

Juan-José Sáenz-Peñafiel[1], Jose-Luis Poza-Lujan[2]([⊠]), and Juan-Luis Posadas-Yagüe[2]

[1] Doctoral School, Universitat Politècnica de València (UPV),
Camino de Vera, s/n., 46022 Valencia, Spain
juasaepe@posgrado.upv.es
[2] University Institute of Control Systems and Industrial Computing (ai2),
Universitat Politècnica de València (UPV),
Camino de Vera, s/n., 46022 Valencia, Spain
{jopolu,jposadas}@ai2.upv.es

Abstract. In recent years there has been a substantial increase in the number of outdoor lighting installations, the energy management of this has not been greatly improved and electricity consumption has skyrocketed. Most of it does not come from renewable energies with all the negative effects that this entails. With all this, public lighting can represent up to a total of 54% of the energy consumption of a municipality and up to 61% of its electricity consumption. This work focuses on the analysis of the factors to consider in the implementation and application of a lighting control system in a real environment for energy saving. The system should be based on the collection of data by the different sensors installed in the luminaries of the route oriented to the environment of the Smart Cities and the Intelligent Transport Systems (ITS). The main objective is to try to reduce the consumption of electrical energy as much as possible while maintaining the comfort that the road user feels in it. For this, the weak points of these systems will be searched and their elimination will be sought. A study will be made of the situation of the systems available today. The characteristics of these systems will be analysed. Based on the characteristics of the systems analysed, the necessary requirements of the system presented will be determined. The characteristics that will make this project different from the rest will be established. An architecture proposal that seeks to optimise the parameters analysed will be presented.

Keywords: Smart Cities · Distributed systems · Distributed architectures · Taxonomy · Power Consumption

© Springer Nature Switzerland AG 2020
F. Herrera et al. (Eds.): DCAI 2019, AISC 1003, pp. 63–70, 2020.
https://doi.org/10.1007/978-3-030-23887-2_8

1   Introduction: Comfort versus Energy Consumption in Smart Cities

The concept of a smart city is in continuous review [2]. However, most authors agree that an intelligent city is one that integrates technology into the elements, and this technology is used to improve the quality of life of the city users. Technology is only the base of the system. Smart cities should also take into account actor, initiatives, goals, and the vision of the city [5].

The use of technology in elements covers a wide technology fields: from the use of information, related to Big Data systems [1], to the use of technology embedded in the everyday objects [3]. One of the most used applications of technology for cities is energy sustainability. Energy saving is a commitment for people in general for their interference in the climate changes that the planet presents. That is why companies and researchers seek alternatives that save energy in several areas, among which are lighting on public roads. The authors and developers of these possible solutions seek to save energy without losing safety and light comfort for the users of the roads, called pedestrians, which generates complexity and a high variety of parameters to be evaluated within an alternative. In this article, an analysis of existing solutions is made to strengthen their strengths and seek to improve their weaknesses. As a result of the analysis, an architecture that fits within the concept of smart cities and at the same time complies with security parameters and light comfort is presented. This architecture aims to optimize the problem of energy consumption on roads through the use of computational algorithms. This optimization is made with two minimization goals: (i) uncomfortability of people and (ii) energy consumption.

The objective is: (1) to show that managing the comfort of a walker or vehicle is complex because there are many variables; (2) demonstrate that it is possible to reach a compromise between energy consumption and comfort; (3) demonstrate that it is possible to manage comfort and consumption through an architecture that manages the "Traffic Adapted Light" mode; (4) demonstrate that a system in which each luminary has a lot of intelligence (embedded agents), it is not necessary to centralise, nor do a complete distribution (bus, star) but "intelligent" connections between luminaries.

The next section will characterise the variables that influence the intelligent management of lighting in a city.

2   State of the Art: Parameters to Measure the Characteristics of Vehicle Mobility in Smart Cities

To define the characteristics to be measured within a system, it is important to set a classification, taxonomy or ontology in which the parameters to be measured within the system are specified [15]. In the Fig. 1 the classification of the proposed environment is observed, separating the System, in which one must have considerations of energy consumption, safety for the users of the roads and budget for both construction and use of the system. On the other hand, the

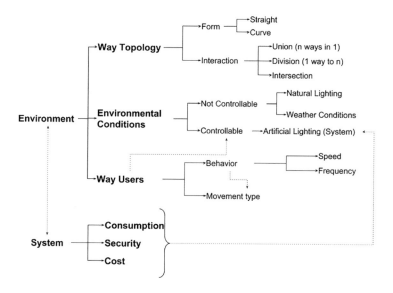

Fig. 1. System characterization.

physical environment in which the system must be implemented, taking into account the topology of the road, i.e. the classification of the roads considering their shape or their interaction between several roads. Within the environment, the lighting conditions are also considered, whether they are controllable by the human (in which the system will be focused), or natural lighting conditions given by nature. The fog conditions on the roads must also be considered, because the amount of lighting necessary to maintain the lighting comfort also depends on this condition. In addition, for the physical environment should be considered the behaviour of users on the roads that are derived by the type of movement and traffic they generate [12].

2.1 Considerations on the Topology of the Tracks, Environment and Pedestrians

For this work has been considered the classification of the routes according to their shape and interaction between them, which gives relevance to the location and lighting of luminaries. For this reason, according to the shape the roads can be classified as straight or as curves according to their physical shape. Regarding the classification by interaction between tracks, it is considered when two or more tracks come together and become one as the union; when a road is divided into two or more roads it is considered as a division and when n roads intersect at a certain point but after crossing they continue as n roads, it is considered as an intersection. For all classifications of the roads, the distance that needs to locate the streetlights, called inter-distance, should be considered. This inter-distance

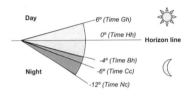

Fig. 2. Crepuscular twilight.

is regulated in the European Union with the standard "EN 13201-2" [16] and can vary from 10 m to 60 depending on the location of the road (Fig. 2).

For lighting on the tracks, the natural environment and therefore the lighting produced by the sun must be considered. For this it is necessary to know the intervals of sunrise and when it is hidden. At these intervals they are called solar twilights. These twilights have been considered for studies of light in the atmosphere, as in [16] where the author specifies the ranges of twilights in relation to human optics. Within this study, the twilights are defined in relation to the distance in degrees of the position of the sun with respect to the horizon that generates the exit and hidden from the sun (2), resulting in three twilights with a difference of 6° to each other [13]:

- Civil Twilight.- 6° with respect to the exit/hidden sun horizon. In this interval, it is not necessary to have artificial lighting.
- Nautical Twilight.- 12° from the exit/hidden horizon of the sun. Natural lighting is still visible to human beings but the European standard for road lighting [16] recommends using artificial light with the lighting conditions of nautical twilight.
- Astronomical Twilight.- 18° from the exit/hidden horizon of the sun. It is necessary to have artificial light during this twilight.

Exceeded the 18° in dawn is considered total natural lighting, while with more than 18° in dusk artificial total lighting is needed. Additionally, we have the blue hours (4° below the horizon) in which the streetlights remain lit and the golden hour (6° above the horizon) which is the start indicator of twilights.

To classify a pedestrian, multiple characteristics (such as speed, acceleration, location in which they are located) must be considered. All these characteristics have been analyzed in studies that seek to classify them by means of their movement and speed [7], determining speed as a priority factor within the characteristics to be analyzed. For this reason, this study uses a pedestrian characterized by a constant speed.

2.2 Considerations About Intelligent Control Systems for Road Environments

Studies and implementations of intelligent control systems for on-street lighting have been carried out, in which the implementation cost is prioritized as in [11] that the company that develops the idea evaluates the cost and benefit of the implemented system and also optimizes compliance with standards. In others the method of capturing the data that will be entered into the system to ensure accuracy and road safety is valued as in [8]. Also, the studies take into account the lighting method required, because this feature helps road safety, in [9] sensorisation is used for ignition, giving a quick response to the walker or vehicle, while other systems as [6] they turn on automatically by schedules, resulting in little adaptability to different environments. In addition, road safety, the adaptability of the luminaires in different environments, the overall energy consumption of the system and the reaction time of the luminaires to be lit should be considered as considered by the system [14]. All these considerations must be met to meet the lighting standards on the tracks. In addition, lighting systems must comply with the quality parameters of systems such as availability, reliability, innocuousness, confidentiality, innocuousness, integrity, maintainability, defined in [4].

3 Proposal

3.1 Explanation of the Scenarios

In this article we propose two ideal scenarios in which we seek to reduce energy consumption without losing the comfort of the walker or vehicle. For this it is considered that the walker or the vehicle needs artificial lighting with the maximum capacity of the luminaries. In addition, it takes a summer day in which the night lasts 9 h between intervals of civil twilight, 8 h in nautical and 6 and a half hours in astronomical. As for the lamp, a lamp with consumption of 30 W/h is used, which is sufficient to meet the European standard. In the scenario, it is assumed that only one bicycle will pass in one direction on the road, at a constant speed of 20 km/h.

3.2 Calculations

To consumption calculate, it is necessary that for the 20 km/h of the bicycle, a braking distance, calculated in [13], of 2 m is needed. Therefore if 2 m is needed for braking, that distance must be illuminated. And if the bicycle has a speed of 5.56 m/s, the lamp must have a minimum ignition time of 0.3597 s, and with safety margin it must remain at least 0.5 s on. To energy consumption calculate, the formula is used:

$$Energy consumed(W) = Consumption(W/h) * time(h) \tag{1}$$

For the calculation of the total energy consumed with an ignition dimmer the formula is used:

$$Total Energy(W) = Energy(W) + Dimmer Energy(W) \tag{2}$$

Table 1. Results of the analysis performed to the scenario presented with a comfort maximum threshold of 2700 Lumens

Method	Consumption (W/day)
Automated Civil Crepuscular	278.1900
Automated Nautical Crepuscular	248.0100
Automated Astronomical Crepuscular	207.0900
Environment Adapted Light	196.0000
Traffic Adapted Light	0.0042
Traffic and Environment Adapted Light	0.0012

Table 1 shows in their rows the different modes analysed. The combination of the different sensors, produce different ways of managing the luminaries. If only one clock is available, the control is automatic (first three lines of the table). If the luminaire has a light sensor, then the mode is an automated adapted light that allows user to have the maximum lumens more time that any clock-based automated mode (fourth row). When luminaries have only sensors to detect walkers or vehicles, the mode is trafic adapted and switch on only light when it is necessary (fifth row). Finally (sixth row) shows the result to combine both sensors: light and traffic. This last mode is the most efficient mode tested. Mode means having greater processing capacity.

3.3 System Proposed

Based on the parameters of the systems that were analyzed and the needs to reduce consumption without losing comfort, a distributed architecture is proposed. The architecture proposed in the Fig. 3 uses sensors of detection of environment for the detection of light and virtual sensors for detection and prediction of traffic. The architecture must allow to obtain data from sensors, process them and act according to the result. For example, with the light sensor it is possible to act on the luminaire to adapt its intensity to the required Lumens intensity by using a dimmer control. Moreover, if the system is implemented by means of a fully distributed system, a luminaire can inform its successors about the presence of a walker or vehicle. The efficient use of a distributed system with intelligent control nodes will allow to complete the proposed objectives.

In this case, the reporting luminaire is considered a source of information or even a virtual sensor. With the data obtained from physical and virtual sensors the outputs will be obtained. These outputs are the events that control the dimmers in order to generate the necessary intensity to the luminaries to provide the needed comfort level for the walker or the vehicle. Environments have topological characteristics that have a great influence on processing. Consequently, the distributed architecture must be capable of supporting a greater intelligence than can be stored in a controller of a luminaire. To support this intelligence, the system must have both advanced control and efficient communications. Therefore,

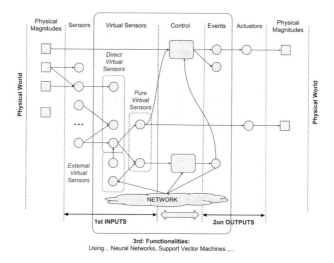

Fig. 3. Proposed architecture.

solutions that cluster the system should be studied, for example, by neighbour-hoods. However, the fact that each zone of action of a luminaire has its own characteristics, suggests that a completely distributed architecture is the most efficient method. For the analysis of input data in the architecture, media techniques, neural vector networks, super vector machines or hybrid multi-objective evolutionary approach techniques that improve performance in sensor networks as recommended by the authors in [10].

4   Conclusions

In this article we present a taxonomy that classifies the characteristics of the luminous control of an intelligent city. The presented taxonomy shows that the intelligent control of the lighting of a city depends on many factors. This high dependence on factors implies that it is necessary to carry out an intelligent control that provides a compromise between the energy consumption and the comfort of the walker or vehicle. Intelligent management must consider both the ambient lighting and the traffic behavior of walker or vehicle. In this way, the energy consumption will be optimized since only lighting is consumed when required. However, the distributed control architecture has a great influence when considering the topology of the tracks. A centralized architecture allows full control of all tracks, however, not all tracks have a similar load. For example, the center of a city will have a different traffic profile than a road connecting two towns. The fact that the topology of the city influences control implies that a centralized architecture can become inefficient in the case of having to control many luminaries. In this case, it should be passed to distributed architectures based on groupings, thus allowing the network and processing load to be adapted to the circumstances of the topology and traffic. In the best case scenario, if each luminaire corresponds to an intelligent node, a fully distributed system based on heterogeneous nodes can provide optimal results. These last points, leave a field of future study very interesting.

Acknowledgments. Work supported by the Spanish Science and Innovation Ministry MICINN: CICYT project PRECON-I4: "Predictable and dependable computer systems for Industry 4.0" TIN2017-86520-C3-1-R.

References

1. Al Nuaimi, E., Al Neyadi, H., Mohamed, N., Al-Jaroodi, J.: Applications of big data to smart cities. J. Internet Serv. Appl. **6**(1), 25 (2015)
2. Albino, V., Berardi, U., Dangelico, R.M.: Smart cities: definitions, dimensions, performance, and initiatives. J. Urban Technol. **22**(1), 3–21 (2015)
3. Arasteh, H., Hosseinnezhad, V., Loia, V., Tommasetti, A., Troisi, O., Shafie-Khah, M., Siano, P.: IoT-based smart cities: a survey. In: 2016 IEEE 16th International Conference on Environment and Electrical Engineering (EEEIC), pp. 1–6. IEEE (2016)
4. Avižienis, A., Laprie, J.C., Randell, B., Landwehr, C.: Basic concepts and taxonomy of dependable and secure computing. IEEE Trans. Dependable Secure Comput. **1**(1), 11–33 (2004)
5. Dameri, R.P.: Searching for smart city definition: a comprehensive proposal. Int. J. Comput. Technol. **11**(5), 2544–2551 (2013)
6. Vadai, T.E., Ziona, N., Shefi, A., Aseret, M., Herbst, E.: (12) United States Patent (10) Patent No .:. 2(12) (2015)
7. Geronimo, D., Lopez, A.M., Sappa, A.D., Graf, T.: Survey of pedestrian detection for advanced driver assistance systems. IEEE Trans. Pattern Anal. Mach. Intell. **32**(7), 1239–1258 (2010)
8. InteliLIGHT: InteliLIGHT® Streetlight Management System (2018)
9. Mahoor, M., Salmasi, F.R., Najafabadi, T.A.: A hierarchical smart street lighting system with brute-force energy optimization. IEEE Sens. J. **17**(9), 2871–2879 (2017)
10. Martins, F.V.C., Carrano, E.G., Wanner, E.F., Takahashi, R.H.C., Mateus, G.R.: A hybrid multiobjective evolutionary approach for improving the performance of wireless sensor networks. IEEE Sens. J. **11**(3), 545–554 (2011)
11. Philips: Smart LED street lighting (2018)
12. Rincon, J.A., Poza-Lujan, J.-L., Julian, V., Posadas-Yagüe, J.-L., Carrascosa, C.: Extending MAM5 meta-model and JaCalIV E framework to integrate smart devices from real environments. PloS One **11**(2), e0149665 (2016)
13. Ruiz, C.: Alumbrado inteligente para el ahorro energético en vías públicas (2018)
14. The Business Case for Smart Street Lights, Silver Spring, p. 6 (2013)
15. Spoladore, D., Arlati, S., Carciotti, S., Nolich, M., Sacco, M.: Roomfort: an ontology-based comfort management application for hotels. Electronics **7**(12), 345 (2018)
16. European Standard: En 13201-2. AENOR (2016)

Integrative Biological, Cognitive and Affective Modeling of Caffeine Use on Stress

Rosaline E. de Haan⬤, Minke Blankert⬤,
and Seyed Sahand Mohammadi Ziabari$^{(\boxtimes)}$⬤

Social AI Group, Vrije Universiteit Amsterdam,
De Boelelaan 1105, Amsterdam, The Netherlands
rosalinedehaan@gmail.com, minkeblankert@gmail.com,
sahandmohammadiziabari@gmail.com

Abstract. In this paper a computational model of the effect of caffeine on cortisol and brain activity levels is presented. Using integrative modelling in a temporal-causal network model approach, this paper aims to dynamically model the biological, cognitive and affective processes that are enhanced in the brain while consuming coffee. Firstly, stress stimuli from the individuals' context conduct an (affective) stressful feeling. Therefore, the individual (cognitive) decides to drink coffee that increases its caffeine levels and makes the individual feel more focused wherefore the individual relaxes. Thirdly, (biologically) it is modeled how caffeine intake by drinking coffee reduces stress levels due to feeling more relaxed.

Keywords: Integrative temporal-causal network model · Biological · Affective · Cognitive · Stress · Caffeine

1 Introduction

Stress has become an increasingly studied subject [3]. Reason for this is the increased societal awareness of stress and its possible negative effects, such as impaired work performance which potentially impacts financial health in organizations and legal obligations and increased interest in employees' health and well-being [25]. Stress can occur when individual senses that they might not have the resources of coping with perceived external stressors [19]. Previous research has identified several possible stressors that can cause stress in students. Examples of stressors in student life could be examinations [1], assignments and essays [1] and striving to meet deadlines [24]. Among working populations workload has been identified as a significant source of stress [7]. In this paper the relation of mental stress caused by workload related stress under influence of caffeine intake has been investigated. There are some previous researches have been done in proposing network-oriented modeling [9, 13–16, 28].

2 Underlying Biological and Neurological Principles

Stress has several effects on the brain. Different brain parts play vital roles in processes caused by stress. Among these brain parts are: the amygdala [23], the hippocampus [23] and the prefrontal cortex [23].

© Springer Nature Switzerland AG 2020
F. Herrera et al. (Eds.): DCAI 2019, AISC 1003, pp. 71–78, 2020.
https://doi.org/10.1007/978-3-030-23887-2_9

Cortisol is a hormone related to stress and needed for different functions in humans. Cortisol is regulated by the hypothalamus–pituitary–adrenal (HPA) axis [18].

Cortisol peaks can be measured after awakening [29], exercising [3] and food intake [21]. However, in these basal conditions, cortisol is not necessarily associated with stress. In these primary conditions, cortisol plays an important role in mobilizing energy resources when external demand increases [18]. This means that the HPA-axis and secretion of cortisol is necessary to cope with stressors. However, the HPA-axis and secretion of cortisol are also known for their associations with stress. The HPA-axis is acknowledged as a stress responsive system that is associated with negative affect in humans [4]. Activation of the HPA-axis can cause impaired functioning in the prefrontal cortex (PFC), resulting in deficits in working memory [23].

It is generally known that caffeine exerts its effects by acting as an adenosine receptor antagonist [3, 4].

These brain functions can be altered by caffeine due to caffeine's structure that is similar to adenosine. Therefore, caffeine serves as an A2a and A1 adenosine-receptor antagonist, meaning caffeine is blocking the effects of the inhibiting effects of adenosine in the brain [25]. Research showed that the amounts of caffeine comparable to one cup of coffee is enough to alter brain functions by blocking the adenosine-receptor activation, resulting in enhanced neurotransmission [19]. There are four adenosine-receptor rich areas in the brain where caffeine has a main effect in functionality: hippocampus, cortex, cerebellum and hypothalamus [23]. Research about the effects of caffeine blocking adenosine-receptors in the hypothalamus have shown that caffeine is able to induce arousal and create wakefulness by inhibiting adenosine' inhibiting effects [8]. Thereby, does adenosine enhancement in the striatum stimulates motor activation. Reason therefore is, that blocking the adenosine-receptor leads to inhibition of dopamine D2-receptor activation which is reason for the reinforcing effects of caffeine leading to motor activation [8, 11]. Thereby, findings in mice research support this claim, due to results showing that mice lacking adenosine-receptors are not showing stimulant effects of caffeine [7].

However which mechanisms are responsible for the cause of these psychostimulant effects is uncertain. Firstly, it has been found that adenosine-receptors control acetylcholine inhibition [4]. Secondly, Caffeine has been found to stimulate dopamine and acetylcholine transmission in the PFC, but also for this it is uncertain if this is the cause or the effect of the psychostimulant effects of caffeine [2]. The stimulation of dopamine became less over time, when tolerance for caffeine occurred [2]. Since cholinergic pathways in the hippocampus and prefrontal cortex (PFC) are linked to behavioral arousal, it is suggested that acetylcholine enhancement by caffeine has stimulating effects on behavioral arousal due to increasing levels of acetylcholine [2, 5, 6]. Furthermore, caffeine is known to strongly interact with dopaminergic neurotransmission, which could be the cause of its motor stimulus effects [8].

Finally, research showed that caffeine results in increasing levels of cortisol during mental stress [21]. Meaning that caffeine and mental stress can cause an additive effect on cortisol production [3, 23], suggesting important interactions between caffeine's effect on the central nervous system and the cortisol components in the stress response. The combination of stress and caffeine result in an increase of anxiety-like behavior [27]. Thereby has research shown that the amygdala is involved in this type of anxiety due to elevated cortisol levels and that this alters physical features such as blood pressure [24, 26]. Contrary, when taking into account that caffeine intake increases during periods of stress [20] and that caffeine is able to result in beneficial effects as described before [10], it would be logical to assume that caffeine is able to create a positive effect on subjective mental stress. Furthermore, research suggest that the adenosine receptor ($A_{2A}R$) could be used to alleviate brain function impairment caused by chronic stress [12].

3   The Temporal-Causal Network Model

In the current model Network-Oriented Modelling has been used to integrate the biological, cognitive and affective processes of caffeine intake by drinking coffee into an overall process. This type of modeling serves as a temporal causal network model approach and has been discussed further and in detail in [27].

Table 1. Explanation of the states in the model

X_1	ws_s	World (body) state of stress s
X_2	ss_s	Sensor state of stress s
X_3	ws_s	World state for context c
X_4	ss_c	Sensor state for context c
X_5	srs_s	Sensory representation state of stress s
X_6	srs_c	Sensory representation state of context c
X_7	fs_s	Feeling state for stress s
X_8	ps_s	Preparation state for stress s
X_9	es_s	Execution state (bodily expression) of stress s
X_{10}	goal	Goal for drinking
X_{11}	ps_{in}	Preparation state for drinking
X_{12}	es_{in}	Drink Caffeine
X_{13}	Prefrontal Cortex	Brain part
X_{14}	Blocking Adenosine Receptor	Neurotransmitter/Hormone
X_{15}	Hypothalamus	Brain part
X_{16}	Cortisol	Neurotransmitter/Hormone
X_{17}	Basal Ganglia	Brain part
X_{18}	Amygdala	Brain part
X_{19}	Pituitary	Brain part
X_{20}	Acetylcholine	Neurotransmitter/Hormone
X_{21}	Hippocampus	Brain part
X_{22}	Noradrenaline	Neurotransmitter/Hormone
X_{23}	Dopamine	Neurotransmitter/Hormone

74 R. E. de Haan et al.

The above conceptual representation has been altered into numerical representations using the theory below [27]:

For states the following combination functions $c_Y(...)$ were used, the identity function id(.) for states with impact from only one other state, and for states with multiple impacts the scaled sum function $ssum_λ(...)$ with scaling factor $λ$, and the advanced logistic sum function $alogistic_{σ,τ}(...)$ with steepness $σ$ and threshold $τ$.

$$id(V) = V$$
$$ssum_λ(V_1, ..., V_k) = (V_1 + ... + V_k)/λ$$
$$alogistic_{σ,τ}(V_1, ..., V_k) = [(1/(1 + e^{-σ(V1+ \ ... \ + Vk \ -τ)})) - 1/(1 + e^{στ})] (1 + e^{-στ})$$

4 Example Simulation

Using neuroscientific literature, a model has been created using qualitative empirical information for the brain mechanisms and showed a possible outcome of the processes involved in this. In Fig. 2 an example of the simulation has been depicted based on the computational model.

The time step was $Δt = 1$ and the scaling factors $λ_i$ for the states with more than one incoming connection and the steepness and threshold factor of the states using adaptive advanced logistic functions are also depicted in Table 2.

In Fig. 3, the equilibrium condition is shown. In this simulation the goal is entered as a constant state with a value of 0.4 (Fig. 1). The simulation has been done in the environment implemented in Matlab [17].

Table 2. Connection weights and scaling factors for the example simulation

Connection weight	$ω_1$	$ω_2$	$ω_3$	$ω_4$	$ω_5$	$ω_6$	$ω_7$	$ω_8$	$ω_9$	$ω_{10}$	$ω_{11}$	$ω_{12}$
Value	1	1	1	1	1	1	1	1	1	1	1	0.8
Connection Weight	$ω_{13}$	$ω_{14}$	$ω_{15}$	$ω_{16}$	$ω_{17}$	$ω_{18}$	$ω_{19}$	$ω_{20}$	$ω_{21}$	$ω_{22}$	$ω_{23}$	$ω_{24}$
Value	1	0.95	1	0.1	0.1	1	0.1	-0.7	1	0.8	1	1
Connection Weight	$ω_{25}$	$ω_{26}$	$ω_{27}$	$ω_{28}$	$ω_{29}$	$ω_{30}$	$ω_{31}$	$ω_{32}$	$ω_{33}$	$ω_{34}$	$ω_{35}$	
Value	0.9	0.8	1	0.7	0.9	1	1	0.1/0.3	0.1	-0.7	0.8	

state	X_5	X_8	X_{15}	X_{19}	X_{21}	X_{23}
$λ_i$	2	2.1	2	1.9	1.1	1.1

state	X_{10}
steepness	500
threshold	0.8

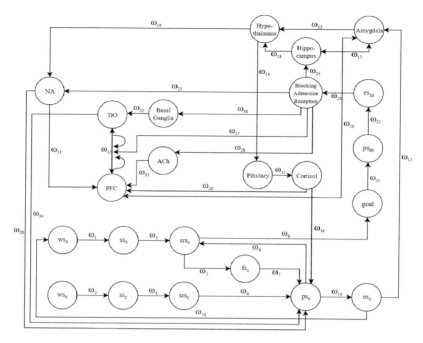

Fig. 1. Conceptual representation of the integrative temporal-causal network model

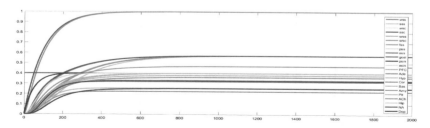

Fig. 2. Simulation results for integrative modeling of the intake of caffeine

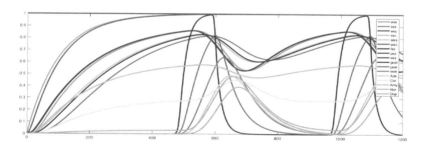

Fig. 3. Simulation result with es_{in} with a constant value of 0.4

5 Mathematical Analysis

In models using the temporal-causal network modelling approach it is necessary to verify the model behavior as expected, using a mathematical analysis method; see [29, Chap. 12].

Using the above explained mathematical analysis was used to obtain the equilibria for the states. For this, linear equations in the equilibrium values of the states were displayed using variables ranging from X1 to X23, for this see Table 1. Using the WIMS Linear Solver[1], mathematical analysis was conducted to obtain the predicted equilibrium values. For this, the following (unique) algebraic solution was used as an input:

$X1 = X9$

$X2 = X1 + 0.1*X23$

$X3 = 1$

$X4 = X3$

$2*X5 = 1*X2 + 1*X8$

$X6 = X4$

$X7 = X5 + 1*X13$

$2.1*X8 = 1*X6 + 1*X7 + 0.1*X16 - 0.7*X22 - 0.7*X23$

$X9 = X8$

$X10 = 0.4 + 0.1*X18$

$X11 = X10 + 0.1*X15$

$X12 = X11$

$2.1*X13 = 0.9*X16 + 1*X18 + 0.1*X20 +$

$X14 = X12$

$2*X15 = 1*X18 + 1*X21$

$X16 = 0.95*X19$

$X17 = 0.8*X14$

$1.9*X18 = 0.8*X9 + 0.1*X21 +$

$X19 = 0.95*X15$

$X20 = 0.7*X14$

$1.1*X21 = 1*X14 +$

$X22 = 0.9*X14 +$

$1.1*X23 = 0.3*X13 + 0.8*X17$

To compare these outcomes with simulation outcomes, in particular the ones depicted in Figs. 2 and 3, the parameter values for $X_3 = 1$ and the value of the X_{10} (goal) = 0.4 were used, because otherwise the other states that are goal dependent are not able to go up and would not reach equilibrium. For verification of these state values found by analysis have been compared with the equilibrium state values found in the simulation. Using a step size of 1 is relatively big (Table 3).

Table 3. Comparing analysis and simulation

State	ws_s X_1	ss_s X_2	ws_c X_3	ss_c X_4	srs_s X_5	srs_c X_6	fs_s X_7	ps_s X_8
Simulation	0,5463	0,5463	1,0000	1,0000	0,5463	1,0000	0,5463	0,5463
Analysis	0,4713	0,4713	1,0000	1,0000	0,4713	1,0000	0,4713	0,4713
Deviation	0,0749	0,0749	0,0000	0,0000	0,0749	0,0000	0,0749	0,0749

---

[1] https://wims.unice.fr/wims/wims.cgi?session=K06C12840B.2&+lang=nl&+module=tool%2Flinear%2Flinsolver.en.

6   Discussion

In this paper an integrative biological, affective and cognitive model has been introduced, where moderate doses of caffeine intake by drinking coffee has been used to lower mental stress levels caused by workload related stressors. This computational model shows that stress influences the goal, preparation state and execution state of drinking coffee and that this results in an increase of blocked adenosine receptors. This influences other hormones and brain area activity related to stress. In our model caffeine intake has been set for moderate levels and the corresponding brain alterations. In the current paper it was hypothesised that caffeine can reduce stress due to the positive and psychostimulant effects caffeine has on individuals.

Furthermore, could future work explore different stressful situations and determine the effects of caffeine on mental stress in those. This is relevant because different kinds of stress have different mental effects. For example, stress in social situations is different from stress related to workload [21].

References

1. Abouserie, R.: Sources and levels of stress in relation to locus of control and self-esteem in university students. Educ. Psychol. **14**, 323–330 (1994)
2. Acquas, E., Tanda, G., Di Chiara, G.: Differential effects of caffeine on dopamine and acetylcholine transmission in brain areas of drug-naive and caffeine-pre-treated rats. Neuropsychopharmacology **27**, 182–193 (2002)
3. Al'Absi, M., Lovallo, W.R., Mckey, B., Sung, B.H., Whitsett, T.L., Wilson, M.F.: Hypothalamic-pituitary-adrenocortical responses to psychological stress and caffeine in men at high and low risk for hypertension. Psychosom. Med. **60**, 521–527 (1998)
4. Al'Absi, M., Lovallo, W.R.: Caffeine's effects on the human stress axis. Coffee Tea Choc. Brain. **2**, 113–131 (2004)
5. Daly, J.W., Fredholm, B.B.: Caffeine - an atypical drug of dependence. Drug Alcohol Depend. **51**, 199–206 (1998)
6. Dua, J.K.: Job stressors and their effects on physical health, emotional health and job satisfaction in a university. J. Educ. Adm. **32**, 59–78 (1994)
7. Ferré, S.: An update on the mechanisms of the psychostimulant effects of caffeine. J. Neurochem. **105**, 1067–1079 (2008)
8. Griffiths, R.R., Evans, S.M., Heishman, S.J., Preston, K.L., Sannerud, C.A., Wolf, B., Woodson, P.P.: Low-dose caffeine discrimination in humans. J. Pharmacol. Exp. Ther. **252**, 970–978 (1990)
9. Lelieveld, I., Storre, G., Mohammadi Ziabari.: A temporal cognitive model of the influence of methylphenidate (Ritalin) on test anxiety. In: Proceedings of the 4th International Congress on Information and Communication Technology (ICICT 2019), 25–26 February. Springer, London, UK (2019)
10. Lovallo, W.R., Farag, N.H., Vincent, A.S., Thomas, T.L., Wilson, M.F.: Cortisol responses to mental stress, exercise, and meals following caffeine intake in men and women (2006)
11. Lovallo, W.R., Whitsett, T.L., Al'Absi, M., Sung, B.H., Vincent, A.S., Wilson, M.F.: Caffeine stimulation of cortisol secretion across the waking hours in relation to caffeine intake level. Psychosom. Med. **67**, 734–739 (2005)

12. McLellan, T.M., Caldwell, J.A., Lieberman, H.R.: A review of caffeine's effects on cognitive, physical and occupational performance. Neurosci. Biobehav. Rev. **71**, 294–312 (2016)
13. Mohammadi Ziabari, S.S., Treur, J.: Computational analysis of gender differences in coping with extreme stressful emotions. In: Proceedings of the 9th International Conference on Biologically Inspired Cognitive Architecture (BICA 2018). Elsevier, Czech Republic (2018)
14. Mohammadi Ziabari, S.S., Treur, J.: Integrative biological, cognitive and affective modeling of a drug-therapy for a post-traumatic stress disorder. In: Proceedings of the 7th International Conference on Theory and Practice of Natural Computing, TPNC 2018. Springer, Berlin (2018)
15. Mohammadi Ziabari, S.S., Treur, J.: An adaptive cognitive temporal-causal network model of a mindfulness therapy based on music. In: Proceedings of the 10th International Conference on Intelligent Human–Computer Interaction, IHCI 2018. Springer, India (2018)
16. Mohammadi Ziabari, S.S.: An adaptive temporal-causal network model for stress extinction using fluoxetine. In: Proceedings of the 15th International Conference on Artificial Intelligence Applications and Innovations (AIAI 2019), 24–26 May. Crete, Greece (2019)
17. Mohammadi Ziabari, S.S., Treur, J.: A modeling environment for dynamic and adaptive network models implemented in Matlab. In: Proceedings of the 4th International Congress on Information and Communication Technology (ICICT 2019), 25–26 Feb
18. Moss, S.E., Lawrence, K.G.: The effects of priming on the self-reporting of perceived stressors and strain. J. Organ. Behav. **18**, 393–403 (1997)
19. Pray, L., Yaktine, A.L., Pankevich, D., Supplements, D., Board, N.: Caffeine in Food and Dietary Supplements: Examining Safety (2014)
20. Ratliff-Crain, J., Kane, J.: Predictors for altering caffeine consumption during stress. Addict. Behav. **20**, 509–516 (1995)
21. Robotham, D., Julian, C.: Stress and the higher education student: a critical review of the literature. J. Further High. Educ. **67**, 734–739 (2005)
22. Roozendaal, B., McEwen, B.S., Chattarji, S.: Stress, memory and the amygdala. Nat. Rev. Neurosci. **10**, 423–433 (2009)
23. Shansky, R.M., Lipps, J.: Stress-induced cognitive dysfunction: hormone-neurotransmitter interactions in the prefrontal cortex. Front. Hum. Neurosci. **7**, 1–6 (2013)
24. Shepard, J.D., Barron, K.W., Myers, D.A.: Corticosterone delivery to the amygdala increases corticotropin-releasing factor mRNA in the central amygdaloid nucleus and anxiety-like behavior. Brain Res. **861**, 288–295 (2000)
25. Smith, B.D., Gupta, U., Gupta, B.S.: Caffeine, mood and performance: a selective review (2007)
26. Smith, J.E., Lawrence, A.D., Diukova, A., Wise, R.G., Rogers, P.J.: Storm in a coffee cup:caffeine modifies brain activation to social signals of threat. Soc. Cogn. Affect. Neurosci. **7**,831–840 (2012)
27. Treur, J.: Network-Oriented Modeling: Addressing Complexity of Cognitive. Affective and Social Interactions. Springer Publishers (2016)
28. Treur, J., Mohammadi Ziabari, S.S.: An adaptive temporal-causal network model for decision making under acute stress. In: Nguyen, N.T. (ed.) Proceedings of the 10th International Conference on Computational Collective Intelligence, ICCCI 2018, vol. 2. Lecture Notes in Computer Science, vol. 11056, pp. 13–25. Springer Publishers, Berlin (2018)
29. Weitzman, E.D., Czeisler, C.A., Zimmerman, J.C., Moore-Ede, M.C.: Biological rhythms in man: relationship of sleep-wake, cortisol, growth hormone, and temperature during temporal isolation. Adv. Biochem. Psychopharmacol. **28**, 475–499 (1981)

Cognitive Modeling of Mindfulness Therapy: Effect of Yoga on Overcoming Stress

Andrei Andrianov, Edoardo Guerriero,
and Seyed Sahand Mohammadi Ziabari[✉]

Social AI Group, Vrije Universiteit Amsterdam, De Boelelaan 1105,
Amsterdam, The Netherlands
Andrei.andrianov@gmail.com,
edoardo.guerriero@gmail.com,
sahandmohammadiziabari@gmail.com

Abstract. Yoga is a practice that is thousands of years old. This practice is also included in the modern mindfulness-based stress reduction training, and its effects on our cognitive system are supported by extensive literature that comprehend mostly fMRI studies and task-oriented experiments with control groups. In this paper, the problem of testing the effects of mindfulness therapy, with specific regard to the yoga practice, is addressed with a Network-Oriented Modelling approach. The first component of the proposed network simulates the elicitation of an extreme stressful emotion due to a strong stress-inducing event. This was done following the same line of previous papers that proposed simulations of similar processes. A second component represents the role of memory, attention and self-awareness in coping with the stressful event. Finally, the yoga practice, divided into physical movements and breathing, is modelled, to show its influence on memory, attention and self-awareness, leading in this way to a reduction of the stress level.

Keywords: Cognitive modeling · Network-Oriented Modelling ·
Temporal-causal network · Mindfulness · Yoga · Stress · Extreme emotions

1 Introduction

In recent years mindfulness-based practices have developed an increased concern as the number of people who pay attention to these techniques and willing to practice them is rising. These techniques were used for the purposes of coping with stress and different clinical applications for a long while now [2–5]. As well as there has been a growing body of literature that recognizes the importance of mindfulness practices in dealing with extreme emotions, and researches which show supporting empirical evidence [9, 10, 18]. We introduce a computational network-based model to simulate the extreme emotional feelings of an individual and the impact of a particular mindfulness technique on these feelings - yoga practice.

Despite the lack of a formalized and shared definition of yoga [22], an important fact to point out is the distinction made by almost all the different yoga schools, between the breathing activity and the training that involves assuming specific body

© Springer Nature Switzerland AG 2020
F. Herrera et al. (Eds.): DCAI 2019, AISC 1003, pp. 79–86, 2020.
https://doi.org/10.1007/978-3-030-23887-2_10

positions. In particular, novice people that are introduced to the yoga practice are always taught to perform specific breathing techniques before to start to practice also the physical training involving the different body positions.

Causal modeling and causal reasoning have a long tradition in science [13–15]. The Network-Oriented Modelling approach based on temporal-causal networks as described in [8, 12] could be considered a continuation of this tradition, also incorporating the dynamic and adaptive temporal perspective on mental states by means of cyclic causal connections, describing cognitive behavior patterns, as in [16]. Temporal-causal network models can be described by means of two equal representations: conceptual representation presented graphically by a graph or by an adjacency matrix; numerical representation in the form of mathematical difference and differential equations. There are some other previous network-oriented modeling literatures for decreasing extreme emotion in [24–29].

2 Neurological Findings

The states of the proposed network model and the connections between them are justified based on the neuropsychological findings from previously literature researches. In this section we discuss these findings, and show the supporting evidence, reported also in the Table 1. There are also some temporal-causal network-oriented modeling literatures about therapies to decrease the stress [6]. It is worth stressing that given that yoga requires performing physical activity, most studies of its effects are carried out with experimental and control groups [11]. The experimental group is usually composed of people who did not practice yoga before, and changes due to this practice are measured after a period of training with questionnaires and sometimes with physiological measurements. Brain imaging techniques are also used to check for significant changes in the brain activity due to yoga practice, but only in longitudinal studies, in particular functional MRI is used mostly to study which brain areas are activated while performing other type of meditation rather than yoga. Given the above, we did not managed to find specific empirical data like fMRI data of subjects recorded while they were practicing yoga, so we based our simulation on generic findings like "Practicing meditation or yoga associated with lower right amygdala volume". Then we link these findings with specific nodes in our network by making use of functional neurological descriptions of the particular brain areas and their interactions.

In general, the literature on mindfulness [1] focus on the benefit that mindfulness can provide in coping with stress and negative emotions. The common sense suggests that general meditations activities (including yoga) should increase the connection with a sort of 'inner self', and suppress negative emotions by providing a detached point of view about ourselves and the experiences we are living. Researches about mindfulness and yoga therefore focused on one of these aspects [7], for example trying to find significant differences in stress perception between people who practice mindfulness and people who don't [21]. Other line of research tried instead to better formalize the role of self-awareness in meditation, trying to differentiate different kind of self-reference, specifically a form of self-awareness related to the present, mainly experienced while performing some meditation activity, and a form of self-awareness more related to a narrative description of the self over time, that is the common self-awareness experienced in our everyday life [18]. Common sense suggests also that performing meditation should help in preventing the degeneration of cognitive functions, and some researchers tried to find significant difference for example in brain areas related to memory between people who are into yoga and other mindfulness practices and people who aren't [17, 20]. Mindfulness practice also seems to increase the performance in episodic and working memory related tasks [23]. We tried to put all these findings into a conceptual representation that we will describe in the next section.

Table 1. Literature findings regarding the impact of mindfulness and yoga on the brain activity.

System involved	Brain area involved	Empirical findings	Reference
Introspection (perception of visceral sensations)	Nucleus Tractus solitarius	Cortisol rapidly affects amplitudes of heartbeat-evoked brain potentials.	[19]
Long Term Memory	Hippocampus	MBSR associated with longitudinal increasing of gray matter concentration within the left hippocampus.	[17]
Planning of complex actions	Brodmann area	BA8 activity correlated with dispositional mindfulness, attention and working memory.	[20]
Stress perception	Amygdala	Practicing meditation or yoga associated with lower right amygdala volume.	[21]
Self-awareness VS narrative-self	Medial Prefrontal Cortex (cortical midline regions)	Experiential self-focus associated with pervasive deactivations along rostral subregions of the dorsal mPFC and ventral mPFC.	[18]
Episodic Memory	Hippocampus	Mindful attention can beneficially impact motivation and episodic memory.	[23]

3   Computational Model

In this section the description of the computational model is presented based on the previously mentioned principles, mainly from the perspective of simulation-based sciences and adopting foundations from Neuroscience, Psychology and Social Sciences. This computational model is a temporal-causal network model based on the Network-Oriented Modelling approach as described in [8].

We designed the following network model with 23 nodes and 43 connections between the nodes, where 2 nodes have initial state values and 5 connections are considered to be negative. To combine multiple effects for the nodes, where applicable, scaled sum and adaptive advanced logistic combination functions were used [6]. For states with single impact connection the identity combination function was used. In Table 2 provided brief explanation for the considered states.

Table 2. Explanation of the states in the model

Variable	Abbreviation	Description	Group
X_1	ws_{ee}	World state for extreme emotion ee	Extreme emotion
X_2	ss_{ee}	Sensory state for extreme emotion ee	Extreme emotion
X_3	ws_c	World state for context c	Extreme emotion
X_4	ss_c	Sensory state for context c	Extreme emotion
X_5	srs_{ee}	Sensory representation state for extreme emotion ee	Extreme emotion
X_6	srs_c	Sensory representation state for context c	Extreme emotion
X_7	fs_{ee}	Feeling state for extreme emotion ee	Extreme emotion
X_8	ps_{ee}	Preparation state for extreme emotion ee	Extreme emotion
X_9	es_{ee}	Execution state for extreme emotion ee	Extreme emotion
X_{10}	$goal$	Goal state for yoga practice	Yoga
X_{11}	sws	Self-awareness state	Memory
X_{12}	ss_b	Sensory state of breathing b	Yoga
X_{13}	srs_b	Sensory representation state for breathing b	Yoga
X_{14}	es_b	Execution state of breathing b	Yoga
X_{15}	mem_e	Memory state for episodic memory e	Memory
X_{16}	ins	Body interoception state	Memory
X_{17}	ps_y	Preparation state for yoga y	Yoga
X_{18}	mem_w	Memory state for working memory w	Memory
X_{19}	ss_m	Sensory state for body movement m	Yoga
X_{20}	srs_m	Sensory representation state for body movement m	Yoga
X_{21}	es_m	Execution state for body movement m	Yoga
X_{22}	mem_{lt}	Memory state for long-term memory lt	Memory
X_{23}	bs_p	Belief state for positive belief p	Memory

The graphical representation is shown below on the Fig. 1, at the bottom of the graph an extreme emotion network is placed as adopted from [6]. On top of the extreme emotion network the memory part is placed as derived from the previous research on mindfulness [17, 18]. The memory network is followed then by the goal and body states for yoga movement and breathing. The connections for the states were considered and justified based on the neuropsychological research as mentioned in the previous section of the paper.

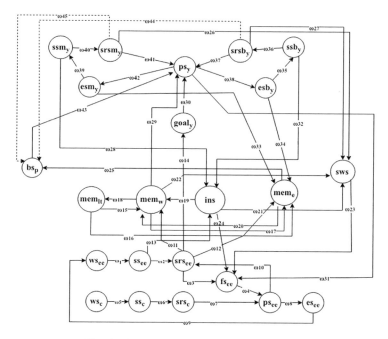

Fig. 1. Graphical representation of the model

4 Mathematical Analysis

This section details the mathematical analysis that has been performed for the model verification. The results are consistent with the simulation experiments as shown in the next section, which confirms the mathematical accuracy of the model. The analysis was performed on the *stationary points* of the model. As defined in [12, Ch. 12], a state Y has a *stationary point* at time t if $\delta Y / \delta t = 0$. The model is in *equilibrium* at time t if every state Y of the model has a stationary point at t.

These criteria were checked for stationary points found by substitution of the values obtained in the simulation and the deviation found. The results of this analysis are shown in Table 3 below. As could be seen from the simulation experiments the model reaches equilibrium around $t = 7000$, thus the verification was done at this time step.

Table 3. Mathematical verification results for stationary points

State	goal	srs_{ee}	mem_w	sws	ins	mem_{lt}	bs_p
Time point	7000	7000	7000	7000	7000	7000	7000
State value	0	0.49	0.307	0.183	0.262	0.221	0.132
Aggregated impact	0	0.541	0.309	0.183	0.27	0.222	0.197
Deviation	0	0.051	0.002	0	0.008	0.001	0.065

5 Simulation Result

In Fig. 3 the resulting plots of the simulation for all the nodes of the network are reported [30]. We already discussed in the previous section the differences found between these plots and our mainly expectations, so in this section we will provide a conceptual interpretation, i.e. how we could interpret the simulation in terms of a real scenario.

The positive believe state increases as well, starting even before than the goal state. This is the hardest state to interpret in a real scenario, since from the literature it is clear that a positive belief towards yoga help people to spend more time into its practice, but it is not so clear how this belief should be represent in a cognitive perspective, and it is also not clear if the belief should push people to practice yoga, or if are the benefits of yoga that make this belief increase over time.

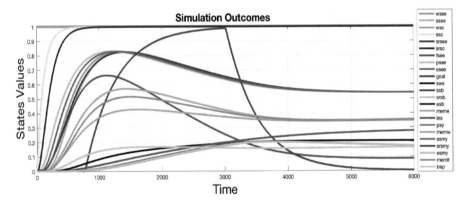

Fig. 3. Simulation results only with plots for all nodes of the network.

6 Discussion

In this paper, we proposed a temporal-causal network that simulate the impact of mindfulness in coping with stressful emotions. The network design was partially based on previous papers on the same topic [6] with the novelty of focusing specifically in the modelling of yoga, including also the role of memory and not just the performing of an action.

Designing the network was not a straightforward process, because of the lack of detailed empirical data on this topic and because of the fuzziness of the definition of some constructs typically involved when talking about mindfulness like *meditations* and *yoga* itself. Nevertheless, the results show that modelling the impact of yoga in coping with stressful emotions is possible, the proposed computational model was verified by the carried mathematical analysis on the stationary points and generally met our expectations which were based on the previous neurological and psychological findings, with some slight deviations. Some adjustments to the proposed network are surely required in order to obtain better and more consistent results.

For future research the proposed model could be extended in several ways. For instance, one could consider adding other specific brain areas involved during in yoga practice, since hopefully in the future more empirical data will be available as well as researches about the neurophysiological basis of mindfulness and yoga in particular. Also, changing could be done to the proposed network design, for example by adding other initial states (the belief node could be an example), or by adding more adaptive connections (for example to suppress one or more state of the extreme emotion group).

In addition to yoga simulation and the previous autogenic training research, other mindfulness-based techniques may be considered for simulation in general, and network-based modeling in particular.

References

1. Bishop, S.R., et al.: Mindfulness: a proposed operational definition. Clin. Psychol. Sci. Pract. **11**, 230–241 (2004)
2. Shapiro, S.L., Schwartz, G.E., Bonner, G.: Effects of mindfulness-based stress reduction on medical and premedical students. J. Behav. Med. **21**(6), 581–599 (1998)
3. Kabat-Zinn, J., et al.: Influence of a mindfulness meditation-based stress reduction intervention on rates of skin clearing in patients with moderate to severe psoriasis undergoing phototherapy (UVB) and photochemotherapy (PUVA) **60**(5), 625–632 (1998)
4. Davidson, R.J., et al.: Alterations in brain and immune function produced by mindfulness meditation. **65**(4), 564–70 (2003)
5. Segal, Z.V., Teasdale, J.D., Williams, J.M., Gemar, M.C.: The mindfulness-based cognitive therapy adherence scale: inter-rater reliability, adherence to protocol and treatment distinctiveness. Clin. Psychol. Psychother. **9**(2), 131–138 (2002)
6. Mohammadi Ziabari, S.S., Treur, J.: Cognitive modelling of mindfulness therapy by autogenic training. In: Proceedings of the 5th International Conference on Information System Design and Intelligent Applications, INDIA 2018. Advances in Intelligent Systems and Computing. Springer, Berlin (2018)
7. Van Gordon, W., Shonin, E., Griffiths, M.D., Singh, N.N.: There is only one mindfulness: why science and Buddhism need to work together. Mindfulness **6**, 49–56 (2014)
8. Treur, J.: Network-Oriented Modeling: Addressing Complexity of Cognitive, Affective and Social Interactions. Springer (2016)
9. Eberth, J., Sedlmeier, P.: The effects of mindfulness meditation: a meta-analysis. Sci. Media **3**(3), 174–189 (2012)
10. Hede, A.: The dynamics of mindfulness in managing emotions and stress. Artic. J. Manag. Dev. **29**(1), 94–110 (2010)
11. Farb Norman, A.S., Anderson, A.K., Mayberg, H., Bean, J., McKeon, D., Segal, Z.V.: Minding one's emotions: mindfulness training alters the neural expression of sadness. Emotion **10**(1), 25–33 (2010)
12. Treur, J.: Verification of temporal-causal network models by mathematical analysis. J. Comput. Sci. **3**(4), 207–221 (2016)
13. Bentler, P.M.: Multivariate analysis with latent variables: causal modeling. Ann. Rev. Psychol. **31**, 419–456 (1980)
14. Kuipers, B., Kassirer, J.P.: Causal reasoning in medicine: analysis of a protocol. Cogn. Sci. **8**, 363–385 (1984)

15. Read, S.J.: Constructing causal scenarios: a knowledge structure approach to causal reasoning. J. Pers. Soc. Psychol. **52**, 288 (1987)
16. Kim, J.: Philosophy of Mind. Westview Press, Boulder (1996)
17. Hölzel, B.K., et al.: Mindfulness practice leads to increases in regional brain gray matter density. Psychiatry Res. **191**(1), 36–43 (2010)
18. Farb Norman, A.S., et al.: Attending to the present: mindfulness meditation reveals distinctneural modes of self-reference. Soc. Cogn. Affect. Neurosci. **2**(4), 313–322 (2007)
19. Schulz, A., Strelzyk, F., Ferreira de Sá, D.S., Naumann, E., Vögele, C., Schächinger, H.: Cortisol rapidly affects amplitudes of heartbeat-evoked brain potentials implications for the contribution of stress to an altered perception of physical sensations? Psychoneuroendocrinology **38**(11), 2686–2693 (2013)
20. Modinos, G., Ormel, J., Aleman, A.: Individual differences in dispositional mindfulness and brain activity involved in reappraisal of emotion. Soc. Cogn. Affective Neurosci. **5**(4), 369–377 (2010)
21. Gotink Rinske, A., et al.: Meditation and yoga practice are associated with smaller right amygdala volume: the Rotterdam study. Brain Imag. Behav. **12**, 1631–1639 (2018)
22. Joshi, K.S.: On the meaning of yoga. Philos. East West **15**(1), 53–64 (2018)
23. Brown, K.W., Goodman, R.J., Ryan, R.M., Anālayo, B.: Mindfulness enhances episodicmemory performance: evidence from a multimethod investigation. PLoS ONE **11**, e0153309 (2016)
24. Mohammadi Ziabari, S.S., Treur, J.: Computational analysis of gender differences in coping with extreme stressful emotions. In: Proceedings of the 9th International Conference on Biologically Inspired Cognitive Architecture (BICA 2018). Elsevier, Czech Republic (2018)
25. Mohammadi Ziabari, S.S., Treur, J.: Integrative Biological, Cognitive and affective modeling of a drug-therapy for a post-traumatic stress disorder. In: Proceedings of the 7th International Conference on theory and practice of natural computing, TPNC 2018. Springer, Berlin (2018)
26. Mohammadi Ziabari, S.S., Treur, J.: An adaptive cognitive temporal-causal network model of a mindfulness therapy based on music. In: Proceedings of the 10th International Conference on Intelligent Human-Computer Interaction, IHCI 2018. Springer, India (2018)
27. Lelieveld, I., Storre, G., Mohammadi Ziabari.: A temporal cognitive model of the influence of methylphenidate (ritalin) on test anxiety. In: Proceedings of the 4th International Congress on Information and Communication Technology (ICICT 2019), 25–26 February. Springer, London, UK (2019)
28. Mohammadi Ziabari, S.S.: An adaptive temporal-causal network model for stress extinction using fluoxetine. In: Proceedings of the 15th International Conference on Artificial Intelligence Applications and Innovations (AIAI 2019), 24–26 May. Crete, Greece (2019)
29. Treur, J., Mohammadi Ziabari, S.S.: An adaptive temporal-causal network model for decision making under acute stress. In: Nguyen, N.T. (ed.). Proceedings of the 10th International Conference on Computational Collective Intelligence, ICCCI 2018, vol. 2. Lecture Notes in Computer Science, vol. 11056, pp. 13–25. Springer Publishers, Berlin (2018)
30. Mohammadi Ziabari, S.S., Treur, J.: A modeling environment for dynamic and adaptive network models implemented in Matlab. In: Proceedings of the 4th International Congress on Information and Communication Technology (ICICT 2019), 25–26 February. Springer, London, UK (2019)

A Healthcare Decision Support System to Overcome Therapeutic Centres Challenges

Bruna Ramos[1(✉)] and João Ramos[2]

[1] Department of Production and Systems, Algoritmi Centre,
University of Minho, Braga, Portugal
bruna.ramos@dps.uminho.pt
[2] Department of Informatics, Algoritmi Centre,
University of Minho, Braga, Portugal
jramos@di.uminho.pt

Abstract. The well-being of a person has been increasingly valued in the present, whether this is related to motor, physical or even psychological development. To help the correct evolution of these main areas there are clinics or therapeutic centres that offer interventions in several areas such as speech therapy, occupational therapy, psychology, psychomotricity, among others. This type of intervention may be performed during a short period of life or even permanently. In both cases it may be important a constant communication between the different professionals in order to provide a harmonious development of the person. This article aims to develop a decision support system for dealing with the time scheduling problem. In addiction, it allows the sharing of information of a patient by the different professionals that are involved. Indeed, there is a promotion of a save sharing environment which leads to global positive results. Due to this type of data, ethical issues and data safety are also considered.

Keywords: Therapy constraints ·
Healthcare decision support system · Time management ·
Time scheduling · Artificial intelligence · Optimization

1 Introduction

Nowadays, issues related to health, physical and psychological development are considered to assess the well-being level of a person. Indeed, there is an increasing concern about these problems. To promote this type of support there are clinics and dedicated therapeutic centres that provide support in the most varied areas, such as speech therapy, psychology, cognitive behavioural therapy, occupational therapy, psychomotricity, physiotherapy, among others. This type of support may be needed by any type of person in different periods of life or

F. Herrera et al. (Eds.): DCAI 2019, AISC 1003, pp. 87–95, 2020.
https://doi.org/10.1007/978-3-030-23887-2_11

even permanently. The incidence of this type of support depends on a large variety of factors and may be performed for different purposes. When this type of support occurs sporadically or for a relatively short time it is considered that the person needs rehabilitation assistance. This may occur already in adulthood, for example, due to an injury requiring specialized treatment in physical therapy. During the age of speech development in a child, speech therapist support is relatively common to ensure correct language development. There are cases in which therapy can be considered as maintenance, meaning that it is necessary for prolonged periods of life or even permanently. These situations may occur due to diagnosed deficiencies or disorders such as cerebral palsy, autism, sensory or behavioural disorders, among others. This type of support may require intervention in several areas and can sometimes be restricted by the availability of the caregiver of the person who needs permanent support.

This work proposes a healthcare decision support system to the time management. The main concern is not only the creation of work schedules, but also the quality of the communication established between the different entities of the system. Thus, we consider that the system will be used by professionals, patients and caregivers. To efficiently manage the available time, it is necessary to understand the main constraints associated with this time of problem. This also includes the establishment of a good communication channel between the different entities, allowing a more efficient time management and information sharing. Another objective of this study is to understand what kind of technology can help to improve the well being of people who attend this type of therapeutic centres.

The remaining of the paper is organized as follows: in Sect. 2 a literature review is performed considered projects related to the research topic; Sect. 3 presents the main challenges that are associated to therapeutic centres; details of the proposed system are described in Sect. 4; finally, in Sect. 5 the main conclusions are drawn and lines for future work are defined.

2   Literature Review

Research in health has been an active topic over last years [16]. Indeed, there are multiple investigation lines and, as a result, the overall quality of live has been improved over the years. Technology has been increasingly used in health and in wellness systems with the intention of improving people's quality of life [7,11,13]. Despite these efforts, there is a need for harmony between the users and professionals who use the system and the technology itself. Only in this way is possible a positive technological change. Several authors have demonstrated their concern in the acceptance of technology [4,8]. This concern occurs due to the lack of acceptance by both patients and professionals [2,12,15]. To become a fully working system, it is necessary to involve the intervening people so that they can be motivated and demonstrate proactivity. In recent years, the use of external devices such as mobile phones, tablets, smartwatches and robots [1,3,14] has been frequent, helping to interact and communicate with the user, collect data and create alert systems. According to Hinton [6], artificial intelligence techniques have potential in health and wellness systems. These techniques are allied to systems that handle large amounts of information and often take advantage of external devices. Another important area that can be allied to artificial intelligence is the use of optimization models allowing a more efficient management of resources and time. According to Fikar et al. [5] there are several challenging optimization problems in the context of healthcare and wellness related to the scheduling and routing problems. Several authors studied the problem that tries to assign the most appropriate nurse to each patient, considering the professional skills and the treatment that the patient needs.

The main objective of this work is to understand the technological challenges associated with a therapy centre. Thus, it is possible to develop a decision support system which consider different specifications, namely facilitate time management; communication between professionals and patients; and the interaction of professionals with emerging technologies. Another future challenge is the development and use of external devices that can somehow improve the development and well-being of patients.

3   Therapeutic Centre Challenges

A therapeutic centre or clinic is an entity that groups a series of specialities of intervention with the aim of promoting physical, motor, psychological and emotional well-being. These entities work in major areas such as speech therapy, occupational therapy, psychology, psychomotricity, physiotherapy, among others.

This article focuses on the particular case of a therapeutic centre that works with children and young adults who essentially require maintenance intervention. Due to the young age or difficulties presented by patients, it becomes essential the constant monitoring by a caregiver. Thus, at the beginning of the school year, after a holiday break, this centre often deals with the problem of creating new schedules. It also deals with cases that may arise after the scheduling period has occurred. This way it is necessary to include a new intervention in the schedule without changing the interventions already established. Another relevant aspect at the time of the creation of the schedules is to perceive where the attendances are realized. These can be performed within the therapeutic centre or in another entity. In the first case the availability of rooms that are attributed to a particular speciality is used. In the second case it is necessary to check the availability of the entity, as well as to recalculate the time required for the intervention and the professionals transportation. These and other restrictive aspects will be presented in Sect. 3.1.

In maintenance therapy it is essential to have a constant communication between all the participants in the process in order to guarantee the success of the intervention. It is essential a good communication between the various professionals of the therapeutic centre who work in different areas. The caregiver and other entities such as schools, other clinics, extra curricular activities, among others stakeholders should be able to access strategies, objectives and progress.

Only in this way there may be cooperation and joint effort to overcome the challenges that arise. This topic will be discussed in more detail in Sect. 3.2.

Another challenge is the use and development of technology, that may or may not use external devices, and improve the quality of life of patients and caregivers. This article focuses more on the problem of scheduling time and communication, but it will certainly be a starting point for the development of technology that promotes the social well-being of people who needs help from therapeutic centres. In designing the architecture of the decision support system this possibility is already taken into account through Sect. 4.

3.1 Time Scheduling Problem

The time management problem in a therapy centre may seem like a simple process, however it deals with a larger number of constraints than the traditional time scheduling problem. Despite dealing with resource constraints common to the traditional time scheduling problem, it adds some time constraints demonstrated by patients and caregivers. The simultaneity of interventions is another important factor that makes time management difficult, since it is necessary to guarantee the presence of certain therapists simultaneously or in consecutive blocks of time. Another important aspect is the temporal management of an intervention in the context of an institution, where there must be a travel time by a therapist or psychologist, for example. This displacement implies a reduction in the working hours of the professional in question. Figure 1 shows the main constraints areas for the proposed time scheduling problem that will be detailed below.

Fig. 1. Main scheduling constraints

Room Availability: There are areas of intervention that require dedicated rooms with specific materials that may not be transported. For example, in an occupational therapy room there must be materials that allow sensory development. In this way certain interventions must be associated with a given room.

However, there may be more than one room available for the same speciality, or even rooms that can be used for specialities that do not require such non-transportable material.

Professional Availability: A professional can work in full-time or in partial-time. In this type of centers it is frequent to find professionals who work only in partial time at the clinic. Another aspect to consider is, for example, pre-established times for meetings, transportation times for professionals to other institutions and other pre-established events.

Simultaneous Availability: Simultaneous availability is a preference when there is collective care or when a caregiver is in charge of one or more patients. This may be the case, for example, of two siblings who have intervention in different areas or even in the same area, but performed by different professionals.

Institution Availability: A particular institution, such as a school, has to authorize intervention in the school context. Despite the authorization, the institution may not allow the occurrence of therapy in certain periods of activities, such as periods that coincide with feeding or other types of routines.

Patient Availability: Patient availability depends not only on temporal availability, but on emotional availability. It is important to evaluate in which part of the day the patient is more predisposed to work. In young children, for example, early in the afternoon may be compromised by sleep state derived from nap time routines.

Routing Time: It is necessary to take into account that when an intervention is performed outside the therapy, for example at a school center, there is an inherent travel time that may restrict the time of intervention of certain professionals.

Caregiver Availability: The availability of caregiver usually depends on the professional status of the person who is delegated the patient. Parents generally have less availability than grandparents or even hired caregivers. That is, this availability varies greatly from patient to patient.

The main goal of the time scheduling problem is to maximize the number of satisfied patients. Due to the hard constraints of the problem it may not be possible to satisfy all requests automatically, since there may be constraints that will have to be relaxed so that the patient can take intervention at the therapeutic centre.

3.2   Communication

Communicating and centralizing clinical data is a sensitive subject that needs an ethical perspective. This subject has been discussed and approached with the constant technological evolution [17]. The advantages and disadvantages of this centralization in several areas have been analysed, but one of the major impacts is in the healthcare area. In [10], the author addresses questions about security and privacy issues with health care information technology. In addition

Fig. 2. Communication process

to informed consent from the patient or caregiver it is necessary to ensure that the data is encrypted and safely stored [9].

Despite the high level of security and consent required, there are many advantages in defining a system that allows constant and current communication between the various members who use the health platform. Figure 2 briefly shows how clinical information can be managed and by whom it should be managed/viewed. For a case of a child in need of maintenance therapy, it becomes important to share information among doctors, therapists, psychologists, educators, caregivers, so that different stakeholders can work in the same direction. It is often difficult to hold periodic meetings to discuss the progress and difficulties presented by the patient. If a child makes rapid progress, constant communication is often needed to define new strategies. In this way it is important to define a system that can support the presented needs. In addition to attendance and the behaviour records during the sessions, it will be possible to define a list of main objectives. Each goal can have an associated progress as well as a list of appropriate strategies to work on different learning and behaviours. This information can be created, updated and discussed among the various professionals who have access to the information. The caregiver may also suggest strategies that will be validated by professionals and share useful information that professionals do not have access to, such as sleep behaviour, behaviour in family or public context, preferred areas, relevant routines, among others. Another advantage of the existence of records and constant communication is the analysis of data. This data may show how a particular patient is evolving and may perceive areas that may have to be worked more intensively or even aspects that no longer need to be worked on so frequently.

4   Decision Support Architecture

The main objective of this work is to define a robust decision support system that takes into account all the particularities related to a therapeutic centre. Figure 3 presents the proposed healthcare system architecture. It is proposed a distributed system where a properly configured server will be in constant communication with the database and the healthcare application. The database contains relevant clinical information that should be protected, so access must be safe. The healthcare application will provide a user-friendly interface that will allow the management of the different contents. This management occurs through a web application that will be available to devices such as the computer, tablet, mobile phone or any other technology that allows access to the internet and a browser. Access to the application can be done by different users who will have access to selected content according to user permissions. Both users and the server can still access external devices such as smart watches, interactive games, among others. Although these important considerations are common to many other traditional applications, the main motivation of this system is the creation of methods that allow the system to be more autonomous and endowed with some technological intelligence.

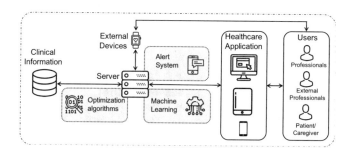

Fig. 3. Healthcare system description

The optimization module integrates a mathematical programming model whose main objective is the maximization of the number of attended patients. This exact approach is based on flow models and job scheduling, including the aforementioned constraints. Due to the inherent complexity of medium and large clinics, a heuristic based on evolutionary computation (e.g., genetic algorithm) is incorporated into this module. This algorithm aims to create a learning process that allows to obtain near optimum solutions in acceptable computational times.

An alert system will be used not only to allow traditional communication, but also to present suggestions or concerns automatically. Through machine learning methods, the system should be able to detect, from the therapists sessions registries, the progress of a user. This may be performed by natural language processing or by using other environmental/wearable sensors (e.g., artificial neural network). According to the performed analysis, the system may alert caregivers and even suggest ways to improve therapy results, for example, by changing the therapy schedule or the location of the therapy (e.g., to the school environment).

5 Conclusions and Future Work

Therapy centres are entities that provide specialized intervention services in order to promote global well-being. These act in several areas such as speech therapy, occupational therapy, psychology, psychomotricity, among others. The main objective of this article is to understand the main challenges in terms of time management, communication and automation of a therapy centre in a technological context. During this article an exhaustive survey is made of requirements that may be essential to the system in order to be able to create a robust decision support system. Another challenge is to understand how the system can be endowed with intelligence through the development of optimization models, the use of natural or evolutionary computation or training of machine learning methods.

Acknowledgements. This work has been supported by FCT - Fundação para a Ciência e a Tecnologia within the Project Scope: UID/CEC/00319/2019.

References

1. Bhavnani, S.P., Narula, J., Sengupta, P.P.: Mobile technology and the digitization of healthcare. Eur. Heart J. **37**(18), 1428–1438 (2016)
2. Chau, P.Y.K., Hu, P.J.H.: Information technology acceptance by individual professionals: a model comparison approach. Decis. Sci. **32**(4), 699–719 (2001)
3. Costa, A., Martinez-Martin, E., Cazorla, M., Julian, V.: PHAROS-PHysical assistant RObot system. Sensors **18**(8), 2633 (2018). https://doi.org/10.3390/s18082633
4. Esmaeilzadeh, P., Sambasivan, M., Kumar, N., Nezakhati, H.: Adoption of technology applications in healthcare: the influence of attitude toward knowledge sharing on technology acceptance in a hospital. In: U- and E-Service, Science and Technology, pp. 17–30. Springer, Heidelberg (2011). https://doi.org/10.1007/978-3-642-27210-3_3
5. Fikar, C., Hirsch, P.: Home health care routing and scheduling: a review. Comput. Oper. Res. **77**, 86–95 (2017). https://doi.org/10.1016/j.cor.2016.07.019
6. Hinton, G.: Deep learning - a technology with the potential to transform health care. JAMA **320**(11), 1101–1102 (2018). https://doi.org/10.1001/jama.2018.11100
7. Jamal, A., McKenzie, K., Clark, M.: The impact of health information technology on the quality of medical and health care: a systematic review. Health Inf. Manag. J. **38**(3), 26–37 (2009). https://doi.org/10.1177/183335830903800305
8. Ketikidis, P., Dimitrovski, T., Lazuras, L., Bath, P.A.: Acceptance of health information technology in health professionals: an application of the revised technology acceptance model. Health Inform. J. **18**(2), 124–134 (2012)
9. Krishna, R., Kelleher, K., Stahlberg, E.: Patient confidentiality in the research use of clinical medical databases. Am. J. Publ. Health **97**(4), 654–658 (2007). https://doi.org/10.2105/AJPH.2006.090902
10. Meingast, M., Roosta, T., Sastry, S.: Security and privacy issues with health care information technology. In: 2006 International Conference of the IEEE Engineering in Medicine and Biology Society, pp. 5453–5458 (2006)

11. Negash, S., Musa, P., Vogel, D., Sahay, S.: Healthcare information technology for development: improvements in people's lives through innovations in the uses of technologies. Inf. Technol. Dev. **24**(2), 189–197 (2018). https://doi.org/10.1080/02681102.2018.1422477
12. Or, C.K.L., Karsh, B.T.: A systematic review of patient acceptance of consumer health information technology. J. Am. Med. Inform. Assoc. **16**(4), 550–560 (2009). https://doi.org/10.1197/jamia.M2888
13. Ossebaard, H.C., Van Gemert-Pijnen, L.: eHealth and quality in health care: implementation time. Int. J. Qual. Health Care **28**(3), 415–419 (2016). https://doi.org/10.1093/intqhc/mzw032
14. Reeder, B., David, A.: Health at hand: a systematic review of smart watch uses for health and wellness. J. Biomed. Inform. **63**, 269–276 (2016)
15. Schaper, L.K., Pervan, G.P.: ICT and OTs: a model of information and communication technology acceptance and utilisation by occupational therapists. Int. J. Med. Inform. **76**, S212–S221 (2007)
16. Thimbleby, H.: Technology and the future of healthcare. J. Publ. Health Res. **2**(3), 28 (2013). https://doi.org/10.4081/jphr.2013.e28
17. Vest, J.R., Gamm, L.D.: Health information exchange: persistent challenges and new strategies. J. Am. Med. Inform. Assoc. **17**(3), 288–294 (2010). https://doi.org/10.1136/jamia.2010.003673

Data Mining, Information Extraction, Semantic, Knowledge Representation

An Intelligent Approach to Design and Development of Personalized Meta Search: Recommendation of Scientific Articles

Jesús Silva[1]($\boxtimes$), Jesús Vargas Villa[2], and Danelys Cabrera[2]

[1] Universidad Peruana de Ciencias Aplicadas, Lima, Peru
jesussilvaUPC@gmail.com
[2] Universidad de la Costa, St. 58 #66, Barranquilla, Atlántico, Colombia
{jvargas41,dcabrera4}@cuc.edu.co

Abstract. In this article we present a method to recommend articles scientists taking into account their degree of generality or specificity. In terms of methodology, two approaches are presented to recommend articles based on Topic Modeling. The first of these is based on the divergence of topics that are given in the documents, while the second is based on the similarity between these topics. After a validation process it was demonstrated that the proposed methods are more efficient than the traditional methods.

Keywords: Information retrieval · Topic modelling · Recommender systems

1 Introduction

The scientific data retrieval, as an activity framed in the discipline of information retrieval, has gained increasing interest in the last times [1]. The impact of the Internet and related technologies has led to the generation of large data sets derived from the actions mentioned above [2]. Likewise, these data led to the development of tools for the management, maintenance, publishing and processing. In this way, different publishers and associations, recognized by part of the scientific community, have published scientific data repositories in websites which constitute important consultation tools for the researcher-user. In addition, there are several tools of this type with a greater or lesser number of features that differ in data source, organization, and the processing performed on them [3]. In this context, a meta-search engine has arisen as an alternative [4] to operate on the area of computer science, to have access to a variety of sources for retrieving and sorting scientific data using an algorithm considering their impact on the scientific community [5]. In this tool, the search for integration of new functionalities and for improvements in both performance and efficiency requires the definition of a homogeneous structure for the storage of the retrieved scientific data, besides the consideration of issues related to the technology to be used, in an operating environment increasingly related to Big Data.

In this article we present an approximation to the recommendation of scientific articles according to their degree of generality or specificity. For this approach, we base

© Springer Nature Switzerland AG 2020
F. Herrera et al. (Eds.): DCAI 2019, AISC 1003, pp. 99–106, 2020.
https://doi.org/10.1007/978-3-030-23887-2_12

ourselves on the notion that the most novice users in a subject, prefer to read as an introduction, while others with a more expert level prefer [6–8].

Our approximations analyze the content of the articles to determine the degree of specificity. Starting from the output provided by a traditional Information Retrieval Systems (RI), our proposal realizes a reordering to offer the most general to specific articles. We use the Latent Dirichlet Allocation (LDA) model proposed by Blei, Ng, and Jordan [9] for topical modeling and we propose two approaches. The first is based on the divergence of topics between articles, while the second is based on the similarity of the same.

2 Theoretical Review

In the recommendation of scientific articles, [10] after analyzing more than 170 algorithms based on Filtering Collaborative (CF) and Content Based (CB), get the best results with those of CF to determine the importance of an article with its citations from the web citations of the Research.net articles. [11], following the idea previous, applies CF algorithms and hybrids to generate an automatic list of articles, evaluating these algorithms on the articles of the ACM Computing Surveys, which give a general overview on a topic, concluding also that the CFs are best in this type of recommendation.

The user profile is one of the most analyzed features to recommend articles. It can be created through the valuation that the user provides to a document, indicating their interest in it [12]. As well as examining the profiles of existing users to detect topics in common among them [13].

To determine the general or specific character of the articles, as far as we know, [14] applies the theory to detect when the topics of the discussion forums are new, general or specific. Following this same line, [15] also applies the theory of sets to topic profiles. These profiles are created by extracting the key words of the title and the first 10 sentences of the original news, including also the comments of the readers.

The approaches described previously determine the general or specific aspects of a news article in the discussion forums. These approaches only apply a topical extraction process based on TF-IDF [16].

Unlike previous jobs. In our approach we obtain a list of articles through the use of a traditional IR system. The LDA model is applied to the content of the articles to extract the topics. These topics are analyzed by their similarity and divergence. And finally we get a reordered list from more general to more specific.

3 Materials and Methods

3.1 Topic Models

There are different types of topic models, among which we can find, Explicit Semantic Analysis, Latent Semantic Analysis, Latent Dirichlet Allocation, Hierarchical Dirichlet

Process, Non-negative Matrix Factorization. As already mentioned in the introduction, we use the Latent Dirichlet Allocation (LDA) [15].

LDA is a probabilistic, generative model of topics. The documents of a corpus are represented as a random combination of latent topics, where each topic is characterized by a distribution of probabilities over a fixed vocabulary of words. For each corpus document Words are generated randomly following the following stages [17]:

1. Randomly select a distribution of topics.
2. For each word in the document:
3. Select a topic randomly on the topics generated in step 1.
4. Select a word randomly over its corresponding vocabulary.

At the end of the process we will obtain the probability of belonging to each word of each document to each topic [18] and [19].

3.2   Description

We will use two methods to analyze the content of the articles, one based on the divergence of the topics of one article and another based on the similarity of topics between articles [15].

To analyze the divergence of the topics we calculate the standard deviation between them, following the intuition that those articles whose deviation is smaller will have a more balanced distribution of topics and therefore have a more general character, since they would not have topics that stand out above the rest. On the other hand, those whose typical deviation is higher would be because there are more representative topics than others in the article, and that therefore it is a more specific article [17].

In the case of the similarity of topics between articles, we will use the similarity of the cosine following the idea that the articles that have a higher average similarity with other articles would be more general, since they have more common topics with other articles. Those that have a low similarity are because they have topics in common with only a few and surely they would be more specific articles. To determine when a scientific article is general or specific. We follow the following steps that are detailed below [20, 21]:

- Consultation: it is the subject that you want to look for in the corpus of scientific articles.
- Indexing and retrieval of articles: we use Lucene, which is a high-performance search engine written in Java and open source. Among its characteristics it allows the incremental indexation of documents, web pages and database contents, among others [22].
  The recovered items are denoted by P = (a1, a2, a3, a4, …, an) and are ordered from highest to lowest relevance. This relevance is given by the default parameters with which Lucene is configured. Therefore, this RI system serves as a first filter to detect the documents most relevant to the study area within a corpus.
  Analysis of the content of the articles: for this we apply the Mallet tool (McCallum, 2002). This tool is based on the LDA algorithm for topical modeling, allowing to detect and extract topics from the corpus of articles. The topics are stored in

different formats and several files as a final result. Our analysis focuses on the file of weights of the topics. This file is made up of a matrix where the articles are in the rows and the weights of the topics of each article in the columns. In other words, each article would be represented by the weights of its topics.

- Results: the list of recommended articles are shown according to their generality (from more general to specific). It can also be configured to show from the most specific to general, that is, according to its specificity.

3.3 Approximation # 1

Divergence In this approach we analyze how the topics differ in the same article. We do this analysis by calculating the standard deviation using Eq. 1 [23]:

$$\sigma(P) = \sqrt{\frac{1}{n-1} \sum_{i=1}^{n} (a_i - \bar{a})^2} \tag{1}$$

As a result, we obtain a weighted value for each article between 0 and 1. These values are interpreted as follows: $\sigma(P) \approx 1$, it is a specific article since the importance of the different topics in the document varies a lot. One another $\sigma(P) \approx 0$, it is a general one since all the topics have approximately the same relevance in the article.

3.4 Approximation # 2

Similarity In this second approach, we analyze the similarity between articles. The similarity is calculated by determining the cosine angle that the topic vectors of the articles form with each other through Eq. 2 [24]:

$$Sim(a_i, a_{i+1}) = \frac{\sum_{k}^{n} a_{ik} \cdot a_{i+1k}}{\sqrt{\sum_{k}^{n} a_{ik}^2} \cdot \sqrt{\sum_{k}^{n} a_{i+1}^2}} \tag{2}$$

As a result we obtain a weighted value for each article between 0 and 1. These values are interpreted as follows: $\sigma(P) \approx 1$, it is a general article since the topics of the article appear in many other articles. $\sigma(P) \approx 0$, it is a specific one since all the topics of the article appear in a few.

3.5 Data

The Scopus database is used, the largest database of citations and abstracts of peer-reviewed bibliographies: scientific journals, books and conference proceedings worldwide. This offers a comprehensive summary of the results of global research in the fields of science, technology, medicine, social sciences and the arts and humanities, Scopus includes intelligent tools to track, analyze and visualize research [25].

4   Results

In order to evaluate the level of generality or specificity of an article, a reference corpus was created, according to Table 1 [26].

Table 1. Scale Reference Corpus

Scale	Description
5	The article provides a general overview of the area (survey), an article that focuses on showing the status of the question for question answering
4	The article is a general overview of a subtopic within the main theme. For example, an article that shows the whole state of the question about the classification of questions applied to question answering systems
3	The article presents a general approach to the area. For example, an article that describes a general approach to solving the question answering problem
2	The article presents an approximation in a domain or concrete area. For example, an article that describes an approximation to the problem of classifying questions in question answering systems, or a general approach to question answering in a specific language.
1	The article presents a specific project or resource. For example, an article that presents a corpus of questions to train question answering systems
0	The article has nothing to do with the area

Two experiments were carried out in order to verify the ideas presented. The objective of these experiments was to determine the optimal number of topographies and evaluate the approximations described above. For the evaluation we used Normalized Discounted Cumulative Gain (nDCG) (see Eq. 3). This measure combines the punctuation of the document (between 0 and 5) with the position in which it has been returned within the list of recommended articles, giving as a single measure the accumulated profit regardless of size. From the list of recovered documents [26].

$$nDCG_P = CR_{P_{[1]}} + \sum_{i=2}^{100} \frac{CR_{P_{[i]}}}{\log_2(i+1)} \tag{3}$$

Where CR refers to the reference corpus.

For experiment 1, the objective was to determine the optimum number of subjects that should be extracted from the corpus of articles to perform their respective analysis of divergence (σ) and similarity (Sim). Topic analysis was calculated nDCG for each set of proposed topics (between 5 and 300), through Eq. 3.

Table 2 shows the nDCG results obtained for 5, 9, 15, 25, 50, 100, 200, 250, 300 and 600 topics, these values were normalized by dividing the DCG result by the maximum possible value obtained when ordering CR. We can see that the best similarity between articles is achieved with 9 topics and the divergence between topics in an article with 600.

In the case of similarity there is a difference of 0.0020 between 9 and 600 topics. Being this difference very small and the greatest value of the mean (M = 0.9345) in the 600 topics. It was decided to fix to 600 topics the optimal value for the second experiment.

With this experiment we conclude that with the analysis of 600 topics per article we have in the first positions the best articles annotated with respect to CRP.

Table 2. Results of nDCG for each group of topics

nDCG	Number of topics									
	5	9	15	25	50	100	200	250	300	600
Sim	0,710	0,720	0,721	0,725	0,732	0,810	0,808	0,812	0,848	0,8469
σ	0,742	0,775	0,812	0,845	0,845	0,852	0,852	0,865	0,886	0,889
M	0,712	0,725	0,732	0,732	0,745	0,752	0,782	0,789	0,812	0,8399

For the second experiment, this experiment aimed to achieve a list of articles where the first ones that recover are the most general ones that address the subject made in the query. With the 600 topics, two approximations were made: the first based on the calculation of the standard deviation for analyze the divergence of the topics in each article, and the second in the cosine similarity to see the co-occurrence of topics between articles.

The study focused on the first 25 articles, applying nDCG @ k, where k = {5, 9, 15, 20 and 25}. This is because from 25 items there is an increase in the nDCG, but this increase is given by the operation of the measure: the reference corpus has lower values in the last positions, even zeros, while our approximation and the experiment reference baseline contain items with a higher rating in those positions.

In this case, we use the nDCG @ k measure, which is the nDCG version that pays more attention to the first results of the list of items retrieved [25].

Table 3 shows the results of applying nDCG @ k to the baseline (Lucene), noting that our approximations in 10, 15, 20 and 25 articles respectively always exceed the baseline.

Table 3. Results of nDGC@k

nDGC@k	Lucene	Sim	σ
5	0,4458	0,3264 (−21,5%)	0,2785 (33,2%)
10	0,552	0,6123 (16,1%)	0,5851 (14,6%)
15	0,5852	0,5910 (7,4%)	0,6954 (22,4%)
20	0,5524	0,6857 (14,7%)	0,6357 (11,0%)
25	0,5857	0,6894 (13,4%)	0,6364 (6,6%)

5 Conclusions

In this article, two approaches based on topic modeling have been presented to recommend scientific articles according to their degree of specificity. First, the divergence of the topics dealt with in each article was analyzed, followed by the similarity of the themes between articles.

By applying topical modeling, the documents returned by an IR system are reordered so that they are displayed in a more general to more specific way.

The first experiment consisted of determining the best possible configuration in terms of the number of topics to be used in the LDA algorithm, obtaining the optimum result with 600 topics.

The second experiment consisted in using our two approaches (based on the divergence of topics in a document and based on the cosine similarity between topics of different documents) to determine the best possible reordering of articles, from more general to more specific.

Our approximations clearly exceeded the baseline using the nDGC measure, obtaining an improvement of 22.4% for nDGC @ 15 using the measure of divergence of topics in a document, and 11% for nDGC @ 20 using the cosine similarity between topics of different documents.

References

1. Agrawal R., Srikant, R.: Fast algorithms for mining association rules in large databases. In: Bocca, J.B., Jarke, M., Zaniolo, C. (eds.) VLDB, Chile, pp. 487–499 (1994)
2. Agarwal, A.K., Badal, N.: A novel approach for intelligent distribution of data warehouses. Egypt. Inform. J. **17**, 147–159 (2015). Elsevier, Egypt (ISSN 1110-8665). https://doi.org/10.1016/j.eij.2015.10.002
3. Agarwal, A.K., Badal, N.: Data storing in intelligent and distributed data warehouse using unique identification number. Int. J. Grid Distrib. Comput. **10**(9), 13–32 (2017). SERSC Australia (ISSN 2005–4262 (Print) ISSN 2207-6379 (Online))
4. Bhaduri, K., Wolf, R., Giannella, C., Kargupta, H.: Distributed decision-tree induction in peer-to-peer systems. Stat. Anal. Data Min. **1**(2), 85–103 (2008)
5. Bose, R., Frew, J.: Lineage retrieval for scientific data processing: a survey. ACM Comput. Surv. CSUR **37**, 1–28 (2005)
6. Butenhof, D.R.: Programming with POSIX Threads. Addison-Wesley Longman Publishing Company, Boston (1997)
7. Chattratichat, J., Darlington, J., Guo, Y., Hedvall, S., Kohler, M., Syed, J.: An architecture for distributed enterprise data mining. In: Proceedings of 7th International Conference on High-Performance Computing and Networking, Netherlands, pp. 573–582 (1999)
8. Chiang, D., Lin, C., Chen, M.: The adaptive approach for storage assignment by mining data of warehouse management system for distribution centre's. Enterp. Inf. Syst. **5**(2), 219–234 (2001)
9. Duan, L., Xu, L., Liu, Y., Lee, J.: Cluster-based outlier detection. Ann. Oper. Res. **168**, 151–168 (2009)
10. Garciarena Ucelay, M.J., Villegas, M.P., Cagnina, L., Errecalde, M.L.: Cross domain author profiling task in spanish language: an experimental study. J. Comput. Sci. Technol. **15**(2) (2015)

11. Gaitán-Angulo, M., Díaz J.C., Viloria, A., Lis-Gutiérrez, J.P., Rodríguez-Garnica, P.A.: Bibliometric analysis of social innovation and complexity (Databases Scopus and Dialnet 2007–2017). In: Tan, Y., Shi, Y., Tang, Q. (eds.) Data Mining and Big Data, DMBD 2018. Lecture Notes in Computer Science, vol. 10943. Springer, Cham (2018)
12. Grossman, R.L., Bailey, S.M., Sivakumar, H., Turinsky, A.L.: Papyrus: a system for data mining over local and wide area clusters and super-clusters. In: Proceedings of ACM/IEEE Conference on Supercomputing, Article No. 63 (1999)
13. Parthasarathy, S., Subramonian, R.: Facilitating data mining on a network of workstations. In: Kargupta, H., Chan, P. (eds.) Advances in Distributed and Parallel Knowledge Discovery. AAAI Press, Menlo Park (2000)
14. Prodromidis, A., Chan, P.K., Stolfo, S.J.: Meta learning in distributed data mining systems: issues and approaches. In: Kargupta, H., Chan, P. (eds.) Book on Advances in Distributed and Parallel Knowledge Discovery. AAAI/MIT Press, Cambridge (2000)
15. Savasere, A., Omiecinski, E., Navathe, S.: An efficient algorithm for data mining association rules in large databases. In: Proceedings of 21st Very Large Data Base Conference, pp. 432–444 (1995)
16. Stolfo, S., Prodromidis, A.L., Tselepis, S., Lee, W., Fan, D.W.: JAM: Java agents for meta-learning over distributed databases. In: Proceedings of 3rd International Conference on Knowledge Discovery and Data Mining, pp. 74–81 (1997)
17. Torres-Samuel, M., Vásquez, C.L., Viloria, A., Varela, N., Hernández-Fernandez, L., Portillo-Medina, R.: Analysis of patterns in the university world rankings webometrics, Shanghai, QS and SIR-SCimago: case Latin America. In: Tan, Y., Shi, Y., Tang, Q. (eds.) Data Mining and Big Data, DMBD 2018. Lecture Notes in Computer Science, vol. 10943. Springer, Cham (2018)
18. Torres-Samuel, M., Vásquez, C., Viloria, A., Borrero, T.C., Varela, N., Cabrera, D., Gaitán-Angulo, M., Lis-Gutiérrez, J.-P.: Efficiency analysis of the visibility of Latin American universities and their impact on the ranking web. In: Tan, Y., Shi, Y., Tang, Q. (eds.) Data Mining and Big Data, DMBD 2018. Lecture Notes in Computer Science, vol. 10943. Springer, Cham (2018)
19. Torres-Samuel, M., Vásquez, C., Viloria, A., Lis-Gutiérrez, J.P., Borrero, T.C., Varela, N.: Web visibility profiles of Top100 Latin American universities. In: Tan, Y., Shi, Y., Tang, Q. (eds) Data Mining and Big Data, DMBD 2018. Lecture Notes in Computer Science, vol. 10943. Springer, Cham (2018)
20. Vásquez, C., Torres-Samuel, M., Viloria, A., Borrero, T.C., Varela, N., Lis-Gutiérrez, J.-P., Gaitán-Angulo, M.: Visibility of research in universities: the triad product-researcher-institution. case: Latin American countries. In: Tan, Y., Shi, Y., Tang, Q. (eds.) Data Mining and Big Data, DMBD 2018. Lecture Notes in Computer Science, vol. 10943. Springer, Cham (2018)
21. Wang, L., et al.: G-Hadoop: MapReduce across distributed data centers for data-intensive computing. Future Gener. Comput. Syst. 29(3), 739–750 (2013)
22. Wang, J., Li, Q., Chen, Y., Liu, J., Zhang, C., Lin, Z.: News recommendation in forum-based social media. J. Inf. Sci. 180(24), 4929–4939 (2010)
23. Giugni, M., León, L.: Cluster doc un sistema de recuperación y recomendación de documentos basado en algoritmos de agrupamiento. Telematique 9(2), 13–28 (2011)
24. Blei, D., Carin, L., Dunson, D.: Probabilistic topic models. IEEE Sig. Process. Mag. 27 (6):55–652010
25. Blei, D., Ng, A.Y., Jordan, M.: Latent Dirichlet allocation. J. Mach. Learn. Res. 3, 993–1022 (2003)
26. Hernández, A., Tomas, D., Navarro, B.: Una aproximación a la recomendación de artículos científicos según su grado de especificidad. Procesamiento del Lenguaje Natural 55(1), 25–51 (2015)

Semantic Links Across Distributed Heterogeneous Data

Sarah Jeter[1]([⊠]), Colleen Rock[1], Brett Benyo[1], Aaron Adler[1], Fusun Yaman[1], Michael Beckerle[2], Alice Mulvehill[3], and Robert Hoh[4]

[1] Raytheon BBN Technologies, Cambridge, MA 02138, USA
{sarah.n.jeter,colleen.rock,brett.benyo,aaron.adler,
fusun.yaman}@raytheon.com
[2] TRESYS Technology, LLC, Columbia, MD 21045, USA
mbeckerle@tresys.com
[3] Memory Based Research, LLC, Pittsburgh, PA 15206, USA
aliceM986@gmail.com
[4] Operationally Focused Acquisition, LLC, Locust Grove, VA 22508, USA
ops-acq@comcast.net

Abstract. Data is often distributed using different formats and terminology to describe the same entities. This can make it difficult to query across data, to find duplicates, and to be aware of related resources. One approach to these issues is unifying the data into a shared format, connecting related entities. We created a prototype human aided system that converts data to RDF and links related entities using semantic web technologies. By using federation, entity resolution, and ontology alignment, common ideas are linked and the user can query across distributed data sources. In this paper, we show that by using our system, the time required to meaningfully connect data is significantly decreased compared to manual linking, showing an improvement of 78% with validation and 99% without validation on an inventory dataset with 11 thousand triples.

Keywords: Semantic · Distributed · Ontology · Alignment · Federated · Query

1 Introduction

Data exists across many domains and formats, often describing the same entity in different ways. The Semantic Web has a natural mechanism for linking independent datasets through URI reuse. Unfortunately, this mechanism is hardly ever used in practice because it requires agreement between data sources, an assumption that goes against the grain of the Semantic Web vision, leaving data as distributed, heterogeneous data sources. In order for a user to find results across the entire linked web, they must know the ontology of each data source.

Building a high quality linked web has been the focus of several communities including those that research ontology alignment, entity resolution, and federated

© Springer Nature Switzerland AG 2020
F. Herrera et al. (Eds.): DCAI 2019, AISC 1003, pp. 107–115, 2020.
https://doi.org/10.1007/978-3-030-23887-2_13

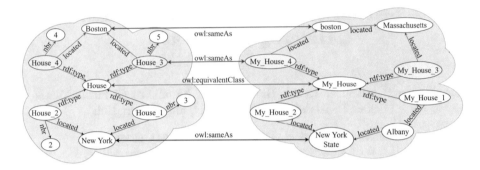

Fig. 1. Alignments and links allow queries to be resolved across multiple graphs. Using the alignments and links between the blue and green graphs (*red lines*), our system allow a query of "SELECT all My_House in Massachusetts with more than 3 bedrooms"

query answering. The Ontology Alignment Evaluation Initiative (OAEI) [1] brings together different technologies across the ontology alignment community each year at a workshop. We have integrated one of the top performers, AML [2], into our Semantic Bridge for aligning classes and properties. Additional aligners that use the Alignment API [3] could be easily integrated. For entity resolution, many systems [4,5] require CSVs as input making DUKE [6] a better choice for our work. Systems that support federated querying often rely on predefined mappings [7] or do not use entity resolution [8]. However, none of these systems integrate both automatic ontology alignment and entity resolution with federated query answering as our system does. We unify the data into a shared format, connecting related entities using ontology alignment to create alignments and entity resolution to create links, creating graphs similar to Fig. 1. With our system, one can easily query over these graphs to find all houses of type "My_House" in Massachusetts with more than three bedrooms, highlighted in yellow, without knowing the ontology of the other graph.

In the rest of this paper, we describe our architecture (Sect. 2), analyze our system (Sect. 3) and conclude (Sect. 4).

2   Architecture

Our system builds upon results from research in ontology alignment and entity resolution to build a linked web in an automated way. The architecture of our system, shown in Fig. 2, is comprised of two types of components, adaptors and the connector. The adaptors convert the data to the *Resource Description Framework* (RDF), while the connector links the RDF. The linked data can then be queried through the Access Manager, a subcomponent of the connector.

2.1   Adaptors

In our design, each adaptor has two main functions: data integration and data analysis. While the implementation of the data analyzer operates over RDF and is the same for every adaptor, the data integrator is specific to a data source.

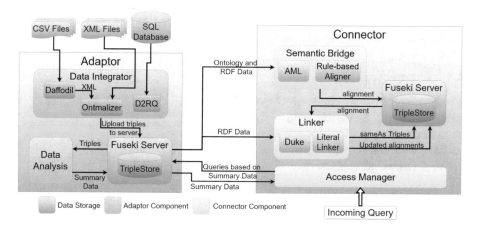

Fig. 2. Architecture of the system demonstrating data flow between components

In its current form, the adaptors expect static data; however, future work could handle data as it is updated by intelligently invalidating and adding triples.

Data Integration. Data integrators utilize configurable parsers and standards-based mapping tools to generate an ontological equivalent of a data source, converting data to RDF. Our system currently supports data stored in XML, CSV, relational databases, and other formats with well-defined schemas.

XML files are converted using Ontmalizer, an open-source tool that converts XSD schemas to ontologies and XML data to RDF [9]. We modified the most recent version to handle imports of XSD Schemas and use the XSD structure to create URIs for custom types [10]. This modified version acts as the nucleus of the system's Ontmalizer Adaptor. Additionally, this adaptor can be configured to use Daffodil [11], an open source implementation of the *Data Format Description Language* (DFDL) [12], to parse data to XML, and convert the XML to RDF using Ontmalizer. If we have the proper DFDL schema, we can parse CSVs this way. We also have an alternative CSV parser if a schema hasn't been created.

The relational database adaptor uses D2RQ [13] to convert data from a relational database to RDF. We built upon the existing D2RQ software to upload the resulting RDF to a Fuseki endpoint [14] and have tested with both MySQL and Derby databases. For data already in RDF format, there is a simple adaptor that will upload it to a local Fuseki instance or connect to a remote endpoint where data already exists and run the data analysis described below.

Data Analysis. We use statistical summaries to get an overview of the linked web for efficient federated query answering. The type of the data (e.g., alphabetical, integer, etc.) is evaluated as well as the count, cardinality, and most frequent values of a property. Additionally, we find the minimum, maximum, and average for each numerical property. These data analysis properties are modeled using the Vocabulary of Interlinked Datasets (VoID) [15]. An example is shown in Fig. 3 where the green data changes with the analysis of each property. These

analysis properties, specifically the subtypes, are used in the rule-base aligner
(see Sect. 2.2) and in the Access Manager to determine where to send queries.

2.2   Connector

A connector is comprised of three
main parts: the Semantic Bridge to
align ontologies, the Linker to resolve
entities, and the Access Manager to
perform federated queries.

Alignment. For non-RDF data sources, adaptors output an ontology
for the raw data as well as convert the data into triple form. *Agreement Maker Light (AML)* is an alignment library that operates over these
ontologies and outputs an alignment
using the Alignment API. For cases
where the data is in triple form with-

```
:graph1
     a void:Dataset ;
     void:propertyPartition :propertyPartition1 .
:propertyPartition1
     void:classPartition :classPartition1 ;
     void:property :exampleProperty1 ;
     sio:count 10 ;
     bbn:cardinality 9 .
:classPartition1
     a void:Dataset ;
     void:class lwap:Integer ;
     dbpedia:average 34 ;
     dbpedia:max 194 ;
     dbpedia:min 0 ;
     void:exampleResource "9" ;
     bbn:most-frequent-value "9" .
```

Fig. 3. Ontology for data analysis, used
in federated query answering and the rule-
based aligner

out an associated ontology, AML would not be able to align the concepts in the
data with the rest of the linked web. To address this case, we designed a rule-
based aligner as an alternative aligning mechanism, which identifies candidates
for alignment and outputs using in the Alignment API. Pair-wise alignments are
created between properties based on subtypes found in the data analysis (i.e.
only a number can be aligned with a number). Designed to be configurable, the
rule-based aligner has rules that can be passed in to the Semantic Bridge to
determine the confidence of an alignment. While there is a set of default rules
that the aligner uses, the user is able to configure the Semantic Bridge with their
own set of rules, allowing the rule-based aligner to be customized for different
datasets.

Entity Resolution. Our Linker has integrated DUKE, an entity resolution engine
that finds duplicate records within a dataset or across multiple datasets. While
DUKE can find equalities between instances, it can not infer similarities between
literals which could represent the same entity. For example, "boeing737" and
"Boeing 737" represent the same aircraft; however, "Boeing 737 NG" and "Boeing 737 MAX" are different. To capture these differences, we designed the Literal Linker to create bbn:sameLiterals objects to link n literals determined to
be similar. Similarity is determined by comparing all objects of two aligned entities with different approaches taken depending on the type of the literal. Only
numbers and strings are currently handled; though, we support multiple string
formats (e.g., alphanumeric vs. only alphabetical) with booleans, anons, and
custom types are handled as strings. For numbers, literals must be exactly the
same, with links created if the types are different (e.g., xsd:int vs. xsd:decimal),
since in RDF the same number with different types are not equivalent to each

other. Strings are compared using an edit distance algorithm and if the string contains a number, it also must be exactly the same. With this, we can tell the difference between "Boeing 737" and "Boeing 747," different aircraft with a high string similarity because only one number is different.

Federation. The Access Manager is based on query decomposition, allowing the user to use a familiar ontology, without needing to know all concepts within the linked web. In Fig. 4, the user is querying for Houses in Massachusetts with more than three bedrooms. Each adaptor knows about part of the query and some properties may use different URIs to mean the same thing (e.g. "nbr" vs. "bedrooms"). A direct query to either adaptor would yield no results, since each has a subset of the needed data.

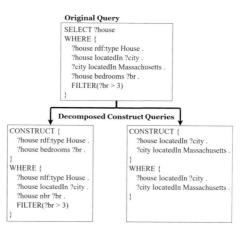

Fig. 4. Example of query decomposition as used in the Access Manager

Queries received by the Access Manager are rewritten into CONSTRUCT subqueries for each adaptor. The WHERE section of each subquery selects statements in the adaptor that could be relevant for answering the original query. For example, the left subquery includes "rdf:type House" and "locatedIn", since the adaptor contains those concepts. The "bedrooms" clause is also included, but rewritten to "nbr", due to the alignment between nbr and bedrooms. Finally, the Massachusetts clause is dropped entirely, since this adaptor has no facts about this entity or anything aligned to it. As each subquery is executed on the different adaptors, the resulting models are combined into an ephemeral model. Since this ephemeral model is in a single ontology due to the rewriting performed by the CONSTRUCT sections, the Access Manager can simply issue the original query on the ephemeral model to get the final result.

3   Analysis

We analyze the performance and accuracy of our system by measuring the time to align and link the data, the accuracy with which we can do so, with and without human verification, and the benefits of using our Access Manager. Table 1 describes the datasets used: two of our own design of inventory

Table 1. Statistics for the following datasets: entity resolution benchmark datasets *(blue)*, Getty Vocab *(red)*, and inventory datasets *(yellow)*

Dataset	Subjects	Predicates	Objects	Edges	#Files	File Type
ERB	79K	123	291K	617K	7	CSV
GV	9,853K	569	16,991K	96,377K	3	RDF
DS 1	2K	275	4K	11K	8	CSV
DS 2	3K	375	6K	15K	8	CSV

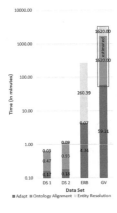
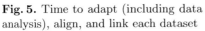

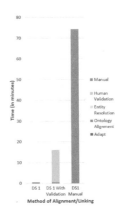

Fig. 5. Time to adapt (including data analysis), align, and link each dataset

Fig. 6. Comparison of methods to align and link data

data, Getty Vocab [16], and entity resolution benchmark datasets [17]. The number of subjects, objects, predicates, and edges for each of the datasets is shown, as well as the number of files it is distributed across and the original file format.

Performance. The time to run the system on a dataset is dependent on the size of the data. Specifically, to convert to RDF and run data analysis, performance is dependent on the number of triples, for alignment, it depends on the number of predicates in each adaptor, and linking depends on the number of found alignments and their objects. A graph of the time to convert, align, and link each of the datasets can be found in Fig. 5. As shown in the graph, smaller datasets, like those we created, can finish within a minute; however, larger datasets may take hours. While this is a significant amount of time, this is a fraction of what it would take to align and link manually. Figure 6 demonstrates the increase in speed that our system provides. Using our system with DS 1, converting, aligning, and linking takes under a minute. With human validation, this increases to 16 min and the entire process manually takes 74 min showing a 78% increase in speed with validation and a 99% increase without when compared to the time to manually link the data.

Alignment Analysis. Just as important as speed, is the accuracy with which we link our data. Within our system, one can choose to use all alignments and links found or a subset that was verified using the user interface (UI); currently a simple developmental interface to interact with the system back end. Each alignment method has their advantages and disadvantages. When only using the alignments and

Table 2. Accuracy analysis for each dataset

Dataset	Alignment Accuracy				Link Accuracy			
	Accuracy	Precision	Recall	F-Score	Accuracy	Precision	Recall	F-Score
ERB	0.658	0.332	0.688	0.444	0.994	0.017	0.483	0.025
DS 1	0.967	0.732	0.963	0.832	0.994	1.000	0.333	0.500
DS 1 w/ Validation	0.998	0.964	0.955	0.959	0.994	1.000	0.333	0.500
DS 2	0.935	0.645	0.897	0.750	0.984	0.708	0.708	0.708

links found by the system, the user risks a lower accuracy, but the analysis is faster with the converse being true for human validation. Also, if the system missed any alignments or links, the user would have to manually verify each possible alignment and link after all. Thus, we have focused on improving recall rather than precision. Table 2 shows the accuracy (1), precision (2), recall (3), and f-score (4) for each dataset, where TP = true positive, TN = true negative, FP = false positive, and FN = false negative. We consistently have high accuracy finding alignments and links automatically which increases with human validation.

$$(1)\ \frac{TP+TN}{TP+TN+FP+FN} \quad (2)\ \frac{TP}{TP+FP} \quad (3)\ \frac{TP}{TP+FN} \quad (4)\ 2*\frac{precision*recall}{precision+recall}$$

Query Federation Analysis. We also analyzed the Access Manager's query federation capability, comparing it to the alternative of importing data from the adaptors directly into a single triple store and issuing a single combined, centralized query. We analyzed three queries, and compared the time from when the query was submitted to the connector, to when the results were ready to present to the user interface (Table 3 Col. 4). The "Altitude" query asked for airports that had a specific type of aircraft available and an altitude greater than a threshold. A variation of this query, "Altitude Restricted," was also created with a higher threshold for the altitude, resulting in less potential matches. The "Availability" query asks for specific details about aircraft, which are stored in a single adaptor. In all of these experiments, the adaptors containing the data were present on the same host as the connector performing the query, in separate processes, with separate triple stores.

Ignoring the cost to ingest data (included in Table 3 Col. 5), the worst case for the federated query, Row 2, was twice as slow as the centralized query. Even in this case, the cost was largely due to the transfer of triples from the adaptor. Including this cost increases the centralized query time by a factor of 10. The "Altitude Restricted" and "Availability" queries show the primary benefit of the federated query approach, where each subquery for the airport dataset only returns the subset of triples corresponding to the more restrictive query. These queries have a reduced data transfer cost due to a restricted number of results and most of the data staying in the adaptors. For example, the last row of Table 3 has the lowest query time as it only requires 6 triples to run the final query. The main difference here is in the federated case, the alignments are checked with

Table 3. Timing metrics (in ms) for federated queries versus centralized queries

Query	Type	Query Size	Query Time	Time w/ Data Ingest	Model Size	Largest Subquery	Final Query
Altitude	Centralized	24 clauses	304 ms	7675 ms	134494 triples	6900 ms	NA
Altitude	Federated	7 clauses	674 ms	674 ms	21322 triples	482 ms	151 ms
Altitude Restricted	Centralized	24 clauses	181 ms	7675 ms	134494 triples	6900 ms	NA
Altitude Restricted	Federated	7 clauses	234 ms	234 ms	3382 triples	155 ms	31 ms
Availability	Centralized	15 clauses	101 ms	7675 ms	134494 triples	6900 ms	NA
Availability	Federated	4 clauses	80 ms	80 ms	6 triples	9 ms	15 ms

optimized java code rather than part of query processing. As the overall data size increases from the smaller examples we have experimented with here to larger multi-million triple datasets, the centralized option can become unfeasible.

4   Conclusions

Distributed heterogeneous data is difficult to use in its present form. By using our system, one is able to increase the speed in which data is converted to RDF, aligned, and linked to other data and easily query across distributed datasets without knowledge of the entire linked web.

Acknowledgment. DISTRIBUTION A. Approved for public release: distribution unlimited. Case Number: 88ABW-2019-0471. This material is based upon work supported by the Air Force Research Laboratory, Rome NY, under Contract No. FA8750-17-C-0193. The authors wish to thank AFRL team members Aaron McVay and Nicholas DelRio.

References

1. Ontology Alignment Evaluation Initiative. http://oaei.ontologymatching.org/. Accessed Dec 2018
2. Faria, D., et al.: The AgreementMakerLight ontology matching system. In: Meersman, R., et al. (eds.) On the Move to Meaningful Internet Systems: OTM 2013 Conferences, OTM 2013. LNCS, vol. 8185. Springer, Berlin, Heidelberg (2013)
3. David, J., et al.: The alignment API 4.0. Semant. web **2**(1), 3–10 (2011)
4. Christen, P.: Febrl: a freely available record linkage system with a graphical user interface. In: Proceedings of HDKM, pp. 17–25. Australian Computer Society, Inc. (2008)
5. Bilenko, M., Mooney, R.J.: Adaptive duplicate detection using learnable string similarity measures. In: Proceedings of ACM SIGKDD, KDD 2003, pp. 39–48. ACM (2003)
6. Garshol, L.M.: Duke. 1.3-SNAPSHOT (2016). https://github.com/larsga/Duke. Accessed Dec 2018
7. Makris, K., et al.: SPARQL-RW: transparent query access over mapped RDF data sources. In: Rundensteiner, E., et al. (eds.) Proceedings of EDBT 2012, pp. 610–613. ACM, New York (2012)
8. Joshi, A.K., et al.: Alignment-based querying of linked open data. In: Meersman, R., et al. (eds.) OTM 2012. LNCS, vol. 7566. Springer, Berlin, Heidelberg (2012)
9. Ontmalizer: Transformation of XML Schemas (XSD) and XML Data to RDF/OWL. SALUS Blog (2013)
10. Rock, C.: crockct/ontmalizer (2018). https://github.com/crockct/ontmalizer. Accessed Dec 2018
11. Apache daffodil (incubating) (2018). https://daffodil.apache.org/. Accessed Dec 2018
12. Beckerle, M.J., Hanson, S.M.: Data format description language (DFDL) v1.0 Specification. Technical report, Open Grid Forum (2014)
13. Bizer, C., Seaborne, A.: D2RQ: treating non-RDF databases as virtual RDF graphs. In: The Semantic Web – ISWC 2004 (2004)

14. Apache Jena Fuseki. https://jena.apache.org/documentation/fuseki2/. Accessed Dec 2018
15. Vocabulary of Interlinked Datasets (VoID). W3C Interest Group Note (2011)
16. The J. Paul Getty Trust: Getty Vocabularies (2018). http://vocab.getty.edu/. Accessed Dec 2018
17. Köpcke, H., Thor, A., Rahm, E.: Evaluation of entity resolution approaches on real-world match problems. Proc. VLDB Endow. **3**(1–2), 484–493 (2010)

Peculiarity Classification of Flat Finishing Motion Based on Tool Trajectory by Using Self-organizing Maps Part 2: Improvement of Clustering Performance Based on Codebook Vector Density

Masaru Teranishi[✉], Shimpei Matsumoto, and Hidetoshi Takeno

Hiroshima Institute of Technology, 2-1-1, Miyake, Hiroshima, Japan
{teranisi,s.matsumoto.gk}@cc.it-hiroshima.ac.jp,
h.takeno.au@it-hiroshima.ac.jp

Abstract. The paper reports an improvement of an unsupervised classification system for learner peculiarities of flat finishing motion of an iron file in skill training. The system classifies and visualizes peculiarities of learners' tool motion effectively by using a torus type Self-Organizing Maps (SOM). An automatic clustering based on codebook density of SOM helps skill trainers to grasp learners' peculiarities distribution easily. In this paper, we focus on the classification improvement of the SOM on the viewpoint of the clustering. Classification performance is improved by comparing different learning schedules of the SOM. Effectiveness of the improvement is evaluated with measured data of an expert and sixteen learners.

Keywords: Motion analysis · Self-organizing maps · Pattern classification

1 Introduction

In the technical education of junior high schools in Japan, new educational tools and materials are in development, for the purpose to transfer kinds of crafting technology. When a learner studies technical skills by using the educational tools, two practices are considered to be important: (1) to imitate experts' motions and (2) to notice their own "peculiarity", and correct it with appropriate aids.

The authors have developed a new technical educational assistant system [1,2] that let learners acquire a flat finishing skill with an iron file, aimed to let learners recognize their own "peculiarity". The system assists the learners how to correct bad peculiarities based on feature of motion that the system extracted. The system classifies the learner's peculiarity features by using a torus type Self-Organizing Maps (SOM) [3,4] effectively. The results of the torus SOM is classified into appropriate number of clusters by automatic clustering algorithm by using "cluster map values" which indicate density information of each

© Springer Nature Switzerland AG 2020
F. Herrera et al. (Eds.): DCAI 2019, AISC 1003, pp. 116–124, 2020.
https://doi.org/10.1007/978-3-030-23887-2_14

codebook vector in the feature space [2]. Skill Trainers could see the learners' peculiarities distribution easily by looking this clustering results. However, the classification method still has room for improvement because our former works [1,2] mainly focused on the developing whole structure of the classification function, therefore the consideration of fine tuning of the core classification mechanisms remained as a future work: for example, optimal selection of learning schedule type of the torus SOM is not yet considered enough.

The paper considers an improvement of classification by comparing different types of learning schedules of torus SOM based on clustering results.

2   Related Work

There are several skill training systems developed with similar concept of our works, for application to various craft skills: lathe operation [5], foundry [6], and brush coating [7]. Those systems provide virtual operation experiences [5,6], or record and playback a learner's operation superimposed on the skilled worker [7] to learners by using virtual reality technology with force feedback devices. From the viewpoint of two important practices described in the introduction, those systems satisfy one of them: (1) imitate experts motions. On the other hand, the rest practice (2) to notice their own "peculiarity", and correct it is not yet implemented none of those systems.

The proposed system has been developed to satisfy both two practices: (1) the force feed back device let learners to imitate skilled worker motion, and (2) motion peculiarity feature extraction based on velocity information [1] and trajectory information [2] present learners to notice their peculiarity easily, and the peculiarity classification method provides trainers to grasp peculiarity distribution of the learners group.

3   Motion Measuring for Flat Finishing Skill Training

Figure 1(a) shows an outlook of the motion measuring devices of the system for flat finishing skill training. The system simulates a flat finishing task that flatten a top surface of an pillar object by an iron file. The system measures a 4D (time series of 3D) motion data of the file by using the PHANTOM Omni (SensAble Technologies) haptics device. The motion is measured as a time series of the position of the file in X, Y, and Z coordinate values, and the posture with the three tilt angles Tx, Ty, Tz as shown in Fig. 1(b). Figure 1(c) shows an example of an expert's motion reciprocating the file 4 times within 20 s. The Y and Z axes correspond vertical and horizontal directions of file along main motion direction of X axis. Since the file works only in the pushing motion, we focused the pushing motion in the classification task in the rest part of the paper.

To classify every file-pushing motion as the same dimensional vector by the SOM, each file-pushing motions parts are clipped out from the measured data, and are re-sampled in 100 time points [1]. Figure 2(b) shows the re-sampled result of the expert motion, Fig. 2(c) is that of a learner. In each plot, all four motions are superimposed.

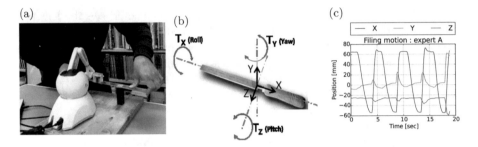

Fig. 1. Flat finishing skill measuring system: (a) outlook, (b) coordinates of measuring, (c) a measured filing motion of an expert.

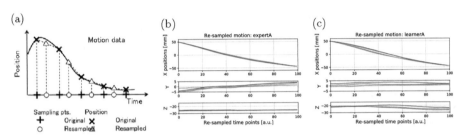

Fig. 2. (a) Re-sampling of a motion, and re-sampled motion:(b) an expert, (c) a learner.

4   Feature Extraction Based on Y-Z Trajectory of Motion

There are relevant differences between the expert's motions and the learner's motions. Firstly, we have focused on main direction of file motion in X coordinate, and have extracted peculiarity feature based on velocity information in our former work [1]. Next we used the Y-Z trajectory as another peculiarity feature as shown in Fig.3(a) in our former work [2] because Y and Z motions are smaller than X motion. The Y-Z trajectories of an expert and a learner are shown in Fig. 3(b) and (c).

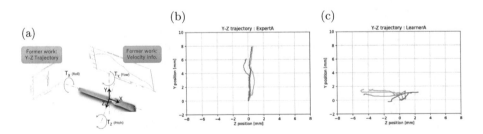

Fig. 3. Y-Z trajectories:(a) concept and examples (b) an expert and (c) a learner.

5   Peculiarity Classification by Torus Type SOM

We classify the file motion peculiarities by using the Self-Organizing Maps (SOM). A technical issue about the classification of the file motion is that the number of peculiarity variations, i.e., the number of classes is unknown. SOM is one of effective classification tools for patterns whose number of classes is unknown. Figure 4(a) shows the structure of the SOM. The SOM is a kind of neural networks which have two layers: one is the input layer, the other is the map layer. The input layer has n neuron units, where n is the dimension of input vector $\boldsymbol{x}$. In this paper, the input vector $\boldsymbol{x}$ consists of a Y-Z trajectory with l re-sampled points, $\boldsymbol{x} = (y_1, y_2, \cdots, y_l, z_1, z_2, \cdots, z_l)^T$, where $n = 2l$. The map layer consists of neuron units, which is arranged in 2D shape. Every unit of the map layer has full connection to all units of the input layer. The ith map unit u_i^{map} has full connection vector $\boldsymbol{m}_i$ which is called "code-book vector". The SOM classifies the input pattern by choosing the "firing code-book" $\boldsymbol{m}_c$ which is the nearest to the input vector $\boldsymbol{x}$ in the meaning of the distance defined as: $\|\boldsymbol{x} - \boldsymbol{m}_c\| = \min_i\{\|\boldsymbol{x} - \boldsymbol{m}_i\|\}$.

   The SOM classifies the high dimensional input pattern vector according to the similarity to the code-book vectors. The map units also arranged in two dimensional grid like shape, and neighbor units have similar code-book vectors. Therefore, the SOM is able to "map" and visualize the distribution of high dimensional input patterns into a simple two dimensional map easily. The SOM constructs the basis of pattern classification by using "Self-Organizing" process with the following three steps onto every input pattern: (1) first, the SOM input a pattern, (2) then it finds a "firing unit" $\boldsymbol{m}_c$, (3) and it modifies code-book vectors of the "firing unit" and its neighbors. In the step (3), code-book vectors are modified toward the input pattern vector. The amount of modification is computed by the following equations, according to a "neighbor function" h_{ci} which is defined based on a distance between each unit and the firing unit. $\boldsymbol{m}_i(t + 1) = \boldsymbol{m}_i(t) + h_{ci}(t)\{\boldsymbol{x}(t) - \boldsymbol{m}_i(t)\}$ where t is the current and $t + 1$ is the next count of the modification iterations. The neighbor function h_{ci} is a function to limit modifications of code-book vectors to local map units which are neighbor the firing unit. We use "Gaussian" type neighbor function. The Gaussian type modifies code-book vectors with varying amounts that decays like Gaussian function, proportional to the distance from the firing unit [3] by $h_{ci} = \alpha(t) \exp\left(-\frac{\|\boldsymbol{r}_c - \boldsymbol{r}_i\|}{2\sigma^2(t)}\right)$, where $\alpha(t)$ is a learning coefficient of modification amount. The standard deviation of the Gaussian function is determined by $\sigma(t)$. We decrease both $\alpha(t)$ and $\sigma(t)$ monotonically as the modification iteration proceeds. The $\boldsymbol{r}_c, \boldsymbol{r}_i$ are the locations of the firing unit and the modified code-book vector unit, respectively. The reason why we use the Gaussian function is based on the assumption that the Y-Z trajectories distribute continuously in the feature space.

Torus Type SOM.   We use the torus type SOM to avoid the "Map edge distortion" problem [2–4]. In the torus SOM, each code-book has cyclic neighbor

Fig. 4. Structure of SOM (a) and torus type SOM (b).

relation in the feature map as shown in Fig. 4(b). In Fig. 4(b), every code-book at the edges of the map adjoins ones at the opposite edges. Since the torus SOM has no map edge, the map is free from edge distortion problem.

6   Clustering Torus SOM Result

The distribution of code-books of the torus type SOM help the skill trainers to look out the peculiarities distribution of learners. We think that a grouping facility of the code-books gives the trainers more efficiency in case of many learners they have. Therefore, the resulted code-books of the torus type SOM are divided into some countable clusters. The proposed system could be more helpful for the trainers by providing such clustering information as types of peculiarities. The trainer could teach according to each peculiarity type effectively in short time. Additionally, such clustering result resolves ambiguous boundary of feature map, which one of disadvantage the torus type SOM has. For this purpose, we introduce automatic clustering method of code-books of the torus type SOM [8].

The clustering method divides code-books into clusters automatically according to densities of the code-books in the feature space. The densities of code-book is named "cluster map". A cluster map value $d(i, j)$ of the code-book $\boldsymbol{m}_{i,j}$ is calculated by

$$d(i, j) = \frac{1}{|D(i, j)|} \sum_{(\mu, \nu)} \in D(i, j)(\boldsymbol{m}_{i,j} - \boldsymbol{m}_{i-\mu, j-\nu})^T \times (\boldsymbol{m}_{i,j} - \boldsymbol{m}_{i-\mu, j-\nu}) \quad (1)$$

where $D(i, j)$ is the first order neighbor region of the map location (i, j), which includes six code-books in the case of hexagonal map topology. The amount of a cluster map value is in inverse proportion to the density of the code-book.

The clustering is done by labeling code-book location (i, j) based on the cluster map by the following algorithm:

(a)  (b)

Fig. 5. (a) Learning schedules and (b) Quantization error.

Step 1: Sort all $d(i,j)$ and index them with numbers $q = 1, 2, \cdots$ in ascending order. Let $s(i,j) = q$ and define $(i,j) = s^{-1}(q)$.
Step 2: Set $L = 0$.
Step 3: Iterate Step 3-1 for $q = 1, 2, \cdots$.
   Step 3-1: if $d(s^{-1}(q))$ is the smallest value among its first order neighbor **then** $L = L + 1$, and assign label $\gamma(s^{-1}(q)) = L$.
   else assign $\gamma(s^{-1}(q))$ with the same label of neighbor which has the smallest d.

7   Improvement of Classification Performance by Different Learning Schedules

The main aim of our former studies [1,2] is to develop whole mechanism of peculiarity classification facility. Therefore, we have not tried to fine tuning of the torus SOM, and also have not evaluated goodness of clustering results yet. The paper mainly aims to improve the torus SOM classification performance by using different learning schedules. The SOM has two types of schedule to decrease learning coefficient $\alpha(t)$: linear type and inverse time type. The linear type schedule reduces $\alpha(t)$ by $\alpha(t) = \alpha_0(1.0 - t/T)$, where α_0 denotes the initial value of $\alpha(t)$ and T is the maximum iteration number. On the other hand, the inverse time schedule reduces $\alpha(t)$ by $\alpha(t) = \alpha_0 C/(C+t)$, where C is a constant that controls the speed of reduction. In this paper, we test three different types of learning schedules: (1) linear type, (2) inverse time type (fast, $C = T/100$), (3) inverse time type (slow, $C = T/10$).

The classification results of those three types of learning schedules are evaluated on the viewpoint of clustering goodness. In this paper, we define the clustering goodness as similarity of classified codebooks in a cluster. If a cluster has almost similar codebooks, the goodness is high. If the cluster has diverse ones, the goodness is low. The clustering goodness is evaluated by observing clustering result manually, and we inspect the relation between the goodness and cluster map values.

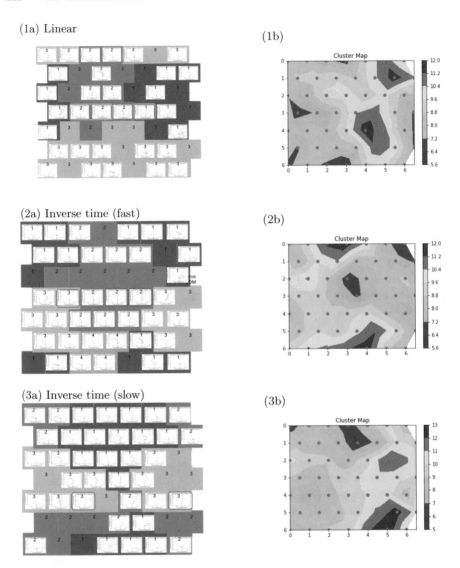

Fig. 6. Clustering result (1a, 2a, 3a) and corresponding cluster map values (1b, 2b, 3b) of torus SOM for each learning schedule. (1) Linear type, (2) Inverse time type (fast), (3) Inverse time type (slow). The expert motions are located the center part of the maps, surrounded by solid lines.

8   Classification Experiment

We carried out experiments of the filing motion peculiarities classification. The motion data were measured with an expert and sixteen learners. Totally we used 66 motion data for classification [1]. We obtain Y-Z trajectories of each re-sampled data with $n = 100$ sampling points. The torus SOM consists the input layer with 200 units and the map layer with 49 code-book vectors, arranged in 7 by 7 with hexa-topology. The each self-organizing process started with initial learning coefficient $\alpha = 0.2$, initial extent of the neighbor function $\sigma = 7$, and iterated 10,000 times with the Gaussian type neighbor function. These types of learning schedules are shown in Fig. 5(a).

The resulted quantization errors for these schedules are shown in Fig. 5(b). All schedules reduced the quantization error in same level at the end of schedules. The Y-Z trajectory peculiarities are clustered as shown in Fig. 6(a) and the corresponding cluster map values are distributed as shown in Fig. 6(b). By comparing Fig. 6(a)s, the (2) inverse time type (fast) presented the highest clustering goodness among the three schedules. The rest both schedules (1) Linear and (3) inverse time (slow) yield low goodness with very diverse codebook vector at cluster 1. On the other hand, by observing cluster map distribution in Fig. 6(2b), the similar cluster map values are distributed within a cluster. Especially, cluster 2 which includes expert motion has totally low cluster map values, which implies the cluster consists of well concentrated codebooks.

The experimental result showed that the different learning schedules yield different clustering result, and we can observe clustering goodness from a point of view. But the concrete calculation of the goodness and clear relation between cluster map and goodness are still remained as future works.

9   Conclusion

The paper proposed an improvement of the peculiarity classification method of flat finishing skill based on Y-Z trajectory of file motions. Three types of learning schedule of the torus SOM are applied, and the inverse type (fast) resulted the best clustering. By comparing the cluster map values of these schedules, good clustering tend to yield similar cluster map value within a cluster.

The concrete formulation of the goodness and clear relation between cluster map and goodness are remained as future works.

The work was supported by JSPS KAKENHI Grant Number 17K04827.

References

1. Teranishi, M., Matsumoto, S., Fujimoto, N., Takeno, H.: Personal peculiarity classification of flat finishing skill training by using torus type self-organizing maps. In: Proceedings of 14th International Conference DCAI 2017, Porto, Portugal, June 2017, pp. 231–238 (2017)
2. Teranishi, M., Matsumoto, S., Fujimoto, N., Takeno, H.: Peculiarity classification of flat finishing motion based on tool trajectory by using self-organizing maps. In: Proceedings of 15th International Conference DCAI 2018, Toledo, Spain, June 2018 (2018)
3. Kohonen, T.: Self-Organizing Maps. Springer, Heidelberg (2001)
4. Ito, M., Miyoshi, T., Masuyama, H.: The characteristics of the torus self-organizing map. In: 6th International Conference on Soft Computing (IIZUKA2000), pp. 239–244 (2000)

5. Li, Z., Qiu, H., Yue, Y.: Development of a learning-training simulator with virtual functions lathe operations. Virtual Real. J. **6**(2), 96–104 (2002)
6. Watanuki, K.: Virtual reality-based job training and human resource development for foundry skilled workers. Int. J. Cast Met. Res. **21**(1–4), 275–280 (2008)
7. Matsumoto, S., Fujimoto, N., Teranishi, M., Takeno, H., Tokuyasu, T.: A brush coating skill training system for manufacturing education at Japanese elementary and junior high schools. Artif. Life Rob. **21**, 69–78 (2016)
8. Tanaka, M., Furukawa, Y., Tanino, T.: Clustering by using self organizing map. J. IEICE **J79-D-II**(2), 301–304 (1996)

An Ontology-Based Deep Learning Approach for Knowledge Graph Completion with Fresh Entities

Elvira Amador-Domínguez[1], Patrick Hohenecker[2], Thomas Lukasiewicz[2], Daniel Manrique[1], and Emilio Serrano[1(✉)]

[1] Department of Artificial Intelligence, Universidad Politécnica de Madrid, Madrid, Spain
{eamador,dmanrique,emilioserra}@fi.upm.es
[2] Department of Computer Science, University of Oxford, Oxford, UK
{patrick.hohenecker,thomas.Lukasiewicz}@cs.ox.ac.uk

Abstract. This paper introduces a new initialization method for knowledge graph (KG) embedding that can leverage ontological information in knowledge graph completion problems, such as link classification and link prediction. Although the initialization method is general and applicable to different KG embedding approaches in the literature, such as TransE or RESCAL, this paper experiments with deep learning and specifically with the neural tensor network (NTN) model. The experimental results show that the proposed method can improve link classification for a given relation by up to 15%. In a second contribution, the proposed method allows for addressing a problem not studied in the literature and introduced here as "KG completion with fresh entities". This is the use of KG embeddings for KG completion when one or several of the entities in a triple (head, relation, tail) has not been observed in the training phase.

Keywords: Statistical relational learning ·
Ontological knowledge base · Knowledge Graph Embedding ·
Latent feature model

1 Introduction

Knowledge representation has always been one of the main challenges of Artificial Intelligence. Throughout time, several approaches have been proposed to model knowledge in a structured and comprehensive way. *Ontologies* [7] are an important approach to knowledge representation. Ontologies present a formal definition of types, properties and relationships between entities applied to a concrete domain. These representations are fairly intuitive for humans, as well as being easily translated into machine languages to allow computers to share concepts. One of the main characteristics of this representation is that concepts or classes are organized in a hierarchical way. These classes tend to remain static, whereas individuals or instances are more dynamic.

© Springer Nature Switzerland AG 2020
F. Herrera et al. (Eds.): DCAI 2019, AISC 1003, pp. 125–133, 2020.
https://doi.org/10.1007/978-3-030-23887-2_15

Knowledge graphs(KGs) are usually supported on this structured form of representation. The approach considered in KGs is, however, different from the one existing in ontologies. Unlike ontologies, KGs aim to gather all the concrete facts existing in the domain instead of extracting general patterns. A fact follows usually the schema (*head, relation, tail*), representing the link between two certain entities by means of a specific relation. These KGs have an increasing popularity in the literature, because they support several tasks, such as question answering or recommendation systems [5].

KGs usually present a high degree of incompleteness, hindering them to be used in a number of domains. Even when completing KGs is mostly a handcrafted task, a number of approaches allow this *Knowledge Graph Completion* (KGC) to be automatic or at least supported automatically [5]. Popular KGC approaches essentially calculate a *Knowledge Graph Embedding* (KGE) that is used for tasks such as: (1) *link or triple prediction*, given two elements of a triple in the KG, ranking the most plausible entities to fulfill the missing element; and (2) *link or triple classification*, predicting if a triple belongs to a KG [10].

Figure 1 illustrates a simplistic KG about American politicians and their relatives. In this graph, Michelle Obama's nationality is unknown. A KGC approach would allow: triple prediction, i.e., ranking the nationalities more likely to fulfill the triple (*Michelle Obama, nationality, ?*); or, triple classification, i.e., calculating the confidence of the triple (*Michelle Obama, nationality, United States*). KGE and KGC approaches could learn that American presidents' wives are likely to be American, although this is not necessarily true as is the case with Donal Trump's wife.

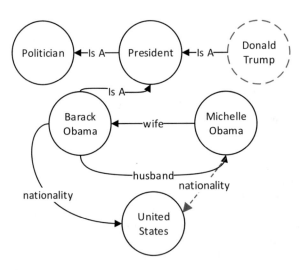

Fig. 1. An example of Knowledge Graph Completion.

Two major shortcomings found in current KGC and KGE proposals are that: (1) they do not leverage ontological information; (2) they are unable to predict facts not seen in training time or *fresh entities*. In Fig. 1, there is not a special treatment for the relation "is A", considered at the same level as relations such as wife or nationality. Furthermore, if the entity Donald Trump is unknown during training, no prediction containing it can be conducted. This paper contributes with a new initialization method for KGEs that addresses these two issues.

The paper outline is as follows. Section 2 presents the related works. Section 3 explains the approach presented. Section 4 details the methodology to evaluate this approach, and Sect. 5 offers the results and discussion. Finally, Sect. 6 concludes and discusses future works.

2   Related Works

2.1   Knowledge Graph Completion

Considering the vastness of the domain they cover, KGs usually present a considerable degree of incompleteness. Therefore, an important challenge in KGs is to complete them automatically. KGC tasks include triple prediction and triple classification. In a more general context, these tasks are in the wider domain of *statistical relational learning*. KGC methods can be categorized into three different groups: (1) graph-feature-based methods, (2) Markov random fields, and (3) latent-feature-based models [5]. In this work, we focus on improving the last group, which includes some of the most popular KGC and KGE models, such as RESCAL and TransE [10].

RESCAL [6] represents a KG as a tensor, where each slice of the tensor represents an existing relation. This method is easy to understand and to implement. On the other hand, the tensor size is increased with the number of entities and relations. Therefore, this approach has scalability problems. TransE, or several of its variants, such as TransD or TransR [10], are more scalable. TransE is based on exploiting distance-based scoring functions. Another of the earliest works in KGC [9] introduces *Neural Tensor Networks* (NTN), an approach based on the combination of the tensor representation with artificial neural networks.

Comparing these approaches is out of the scope of this paper, and there are extensive reviews [5,10]. The approach presented is an intialization method that can be combined with these classical works in KGC and KGEs. In this paper, we have experimented with the NTN model, since its implementation and results are available online, facilitating the evaluation and comparison of our proposal.

2.2   KGC with Ontologies

As noted by Paulheim [8], ontologies are mainly involved in the creation and data introduction phase of KGs. They are useful for this phase, as they provide restrictions for instances and relations.

Ontological information is not leveraged in the mainstream of KGE and KGC approaches [10]. Some works address its use for the *type inference* problem, i.e., predicting the type of a given entity. This can be considered as a KGC task, although less versatile than triple prediction and triple classification. Only Hohenecker and Lukasiewicz [2] recently propose to distinguish ontologies and facts for a number of KGC tasks. Unlike in this previous work, instead of presenting a new KGE approach, a simple initialization method is proposed to be combined with classical KGEs. This method, as shown in Sect. 5, has potential of enhancing KGC via the use of ontologies.

2.3   KGC with Fresh Entities

As explained in the introduction, fresh entities refer to those entities that have not been seen in training time when building a KGE model. Most KGE approaches cannot deal with these entities, because they do not receive an initial encoding in the one-hot vector initialization [5]. Some proposals technically can use fresh entities as input in the prediction phase. In this vein, Socher et al. [9] use Word2Vec [3] embeddings as input in the KGE model. However, beyond having a compatible interface, these works do not explore or evaluate the possibility of reasoning over fresh entities as in the presented contribution.

3   An Ontology-Based Deep Learning Approach for Knowledge Graph Completion with Fresh Entities

The hypothesis presented is that KGC can be enhanced if the initial representation of entities is complemented with a vector that encodes ontological information. In the example given in Fig. 1, predicting links for the entity "Barack Obama", let us call it e_1, should be easier if the initial representation of e_1 is enriched with an embedding that tries to reflect that e_1 is a president and, therefore, a politician. Moreover, "fresh entities" could also receive this extra information about the ontology if their initial representation can deal with entities not seen in training time. In the example, predicting links for "Donald Trump", denoted e_2, would be feasible if, at least, e_2's links with the ontology were known. That is, even when e_2 could be new in prediction time, a KGC approach could reason about it if e_2's initial representation encodes that e_2 is a president and a politician. This is because the information about similar entities such as e_1 might be leveraged.

Figure 2 summarizes the method proposed and its application to general KGs. Firstly, the ontological information or *ontological knowledge base* (OKB), such as concepts and classes, has to be separated from the *general knowledge base* (KB), such as individuals or instances. In the example, only the relation *type_of*, which is present in a number of general KGs such as Freebase, is taken as the hierarchy of concepts in the ontology. Secondly, an initial representation is chosen for the entities in both the OKB and the KB. Following Socher et al.'s [9] initialization method, the mean of vectors using Word2Vec is used for each word contained in an entity. For example, the initial value "Bengal tiger" is the mean of vectors for the words "Bengal" and "tiger". In the case of the OKB, this initial mean of vectors is enriched with the concepts higher up in the hierarchy. For instance, the

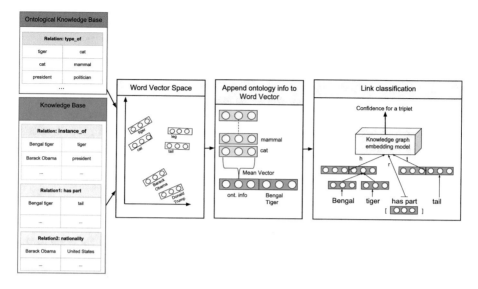

Fig. 2. An ontology-based deep learning approach for KGC with fresh entities

initial value for the concept "tiger" is the mean of vectors for the words "tiger", "cat", and "mammal". Thirdly, each entity vector in the KB is extended with its corresponding OKB vector codifying ontological information. Fourthly, a KGE is learned using known approaches such as NTN or RESCAL. Finally, this KGE can be used to solve KGC tasks such as link classification.

The main idea behind this approach is that relations tend to have a type restriction for the entities that can be linked. These restrictions are explicitly represented in ontologies. The proposed initialization method includes this type information explicitly in the initial representations for entities, releasing KGE approaches from the need to learn it. Moreover, considering an embedding for the hierarchy of types or concepts in the input can allow KGEs to reason about "fresh entities" in certain relations, even if nothing else is known about the new entity. Furthermore, pretrained Word2Vec models can leverage information from unsupervised texts and then transfer this information to the KGC process.

4   Methodology

4.1   Baseline Model

To evaluate the proposed method, a Neural Tensor Network (NTN)[9] is used as KGE approach[1]. One of the main advantages of this model is that both the datasets and the original code are publicly available, allowing an easy comparison of the use of NTN alone or with the proposed initialization method containing

---

[1] The implementation used in this work, as well as the employed datasets are available at https://github.com/Elviish/ntn-pytorch-ontological-info.

ontological information. In this model, the embedding matrix is incorporated as a parameter of the network, allowing it to be initialized externally.

4.2 Datasets

Two different datasets are considered to evaluate our approach. As mentioned, Socher et al. [9] provide not only the model but also the data used for training and evaluation. Moreover, these provided datasets are hegemonic in the literature and widely used to evaluate different KGC approaches. These datasets represent a reduced subset of the data existing in WordNet [4] and Freebase [1], providing facts about 11 and 13 relations, respectively.

4.3 Training Protocol

Before the training phase, a type-hierarchy retrieval is performed for each entity on the dataset as explained in Sect. 3. The DBpedia ontology is the selected source for type retrieval and ontological information. A SPARQL query is run to obtain each entity type or class and its upper classes. Then this set is transformed into a vector and appended to the initial embedding of each entity. The ontological part of the initial entity embedding is frozen throughout the training phase.

The model is trained for the triple classification, a binary classification problem. Therefore, the model has to be fed with both positive and negative examples in order to learn to discriminate. Since the training datasets are only composed by correct triples or positive samples, ten negative samples are generated for each positive example by exchanging the tail entity of the given fact with a randomly sampled one from the total entity pool. The NTN is trained with 500 epochs and a regularization value of 1e-4.

4.4 Evaluation Protocol

Regarding evaluation, two different scenarios are considered. Firstly, the impact of the ontological information on the model is quantified by comparing the results obtained with or without appending the ontological information vector to the entity. In these experiments, the provided test sets composed by the same amount of correct and incorrect facts are used. All facts are formed by entities that have been previously seen during training. However, not all the relations in the datasets are evaluated cause some of them are considered unsuitable for triple prediction even for humans, such as *place of death* in Freebase.

In the second scenario, a new dataset partition is considered to evaluate if the ontology-based initialization method allows reasoning with "fresh entities". The first set is composed by facts where both head and tail entities are known, whereas the second set is composed by facts where either the head or the tail are unknown in training time. In this case, ontological information is always considered. This fresh entities set is generated by randomly selecting 1500 entities from the total set and extracting its associated facts in the training data.

Table 1. Results over WordNet and FreeBase, both considering only the knowledge based without and with ontological information, and results over WordNet and Freebase, both considering all entities in training time or with fresh entities in testing.

Relation	WordNet		WordNet with FEs		Relation	Freebase		Freebase with FEs	
	KB	OKB	No FEs	FEs		KB	OKB	No FEs	FEs
member_meronym	0.51	0.53	0.54	0.52	nationality	0.37	0.5	0.53	0.32
member_holonym	0.52	0.5	0.53	0.53	profession	0.47	0.6	0.5	0.5
part_of	0.52	0.54	0.55	0.53	institution	0.52	0.5	0.56	0.58
has_part	0.52	0.5	0.5	0.48	cause_of_death	0.52	0.53	0.57	0.7
domain_region	0.49	0.56	0.48	0.49	religion	0.39	0.54	0.55	0.55
sysnet_domain_topic	0.44	0.53	0.47	0.47	ethnicity	0.5	0.53	0.36	0.32
domain_topic	0.51	0.57	0.49	0.44					
MEAN	0.51	0.53	0.52	0.51		0.45	0.53	0.53	0.41

5   Results and Discussion

Table 1 shows the results in terms of accuracy for the triple classification problem. As shown, there is an improvement with the proposed method for all relations studied in WordNet except *member_holonym* and *has_part*. For the relation *sysnet_domain_topic* the method produces a 9% enhancement. For Freebase, there is an improvement with the proposed method for all relations studied except in the relation *institution*. For the relation *religion*, there is a 15% increment. The relation *profession* also presents a significant increase, 13%.

The difference between WordNet and Freebase can be motivated cause, while WordNet relations are not restricted by the entity class but by its syntactical category, the entity types in Freebase relations are more homogeneous.

Regarding the fresh entities problem, results show that the approach presented can address it. There is a slight worsening, always less than 5%, in the prediction of three of seven relations evaluated for WordNet. For Freebase, only in two of six relations the accuracy is lower when fresh entities are considered. However, for the specific case of *nationality*, the loss of accuracy is considerably high, 21%.

These results support the hypothesis that the explicit ontological information used in the proposed initialization method allows KGEs to reason about entities not seen in training time by leveraging type restrictions.

6   Conclusion and Future Works

A novel initialization approach for *Knowledge Graph* (KG) completion models is presented. In this approach, explicit ontological information is introduced in the form of a vector that gathers the class hierarchy of a given entity. This information is fixed and shared across similar entities, which leads the KG completion model to infer type restrictions on the relations. This approach not only increases the predictive capability of the model in the canonical *triple classification* problem (up to 15%), but also enables reasoning over entities unknown in training time or *"fresh entities"*. To the best of our knowledge, this is the first research work introducing the fresh entities problem, proposing an approach to address it, and evaluating it.

Since KGs are very dynamic, and obtaining *KG Embeddings* (KGEs) is computationally expensive, addressing the fresh entities problem allow trained KGEs to be used when new entities and links are added in the KG without retraining the model. More importantly, this allow transfer learning with KGEs, i.e., using pretrained KGEs for a number of different KGs as long as they share the same ontology. This is important because, unlike the instances than can be private and very dynamic, ontologies tend to be public, general, and a more static source of information. The presented experimental results show that dealing with this interesting and challenging problem is feasible, although with a cost in the predictive power, as expected.

Our future works include: applying this initialization approach to new knowledge bases; evaluating its use in other KGE methods, such as TransE, RESCAL, or ComplEx [10]; and evaluating transfer learning among KGs sharing the same ontological information.

Acknowledgments. This research work is supported by the Spanish Ministry of Science, Innovation and Universities under the program "Estancias de movilidad en el extranjero José Castillejo para jóvenes doctores" (CAS18/00229) and by the "Universidad Politécnica de Madrid" under the programs: "Ayudas al Personal Docente e Investigador para Estancias Breves en el Extranjero", "Ayudas dirigidas a Jóvenes Investigadores para Fortalecer sus Planes de Investigación", and "Ayudas para Contratos Predoctorales para la Realización del Doctorado".

References

1. Bollacker, K., Evans, C., Paritosh, P., Sturge, T., Taylor, J.: Freebase: a collaboratively created graph database for structuring human knowledge. In: Proceedings of the 2008 ACM SIGMOD International Conference on Management of Data, SIGMOD 2008, pp. 1247–1250. ACM, New York (2008)
2. Hohenecker, P., Lukasiewicz, T.: Ontology reasoning with deep neural networks. CoRR, arxiv: abs/1808.07980 (2018)
3. Mikolov, T., Chen, K., Corrado, G.S., Dean, J.: Efficient estimation of word representations in vector space. CoRR, arxiv: abs/1301.3781 (2013)
4. Miller, G.A.: WordNet: a lexical database for English. Commun. ACM **38**, 39–41 (1995)
5. Nickel, M., Murphy, K., Tresp, V., Gabrilovich, E.: A review of relational machine learning for knowledge graphs: from multi-relational link prediction to automated knowledge graph construction. CoRR, arxiv: abs/1503.00759 (2015)
6. Nickel, M., Tresp, V., Kriegel, H.-P.: A three-way model for collective learning on multi-relational data (2011)
7. Noy, N.F., McGuinness, D.L.: Ontology development 101: a guide to creating your first ontology (2001)

8. Paulheim, H.: Knowledge graph refinement: a survey of approaches and evaluation methods. Semantic Web **8**(3), 489–508 (2017)
9. Socher, R., Chen, D., Manning, C.D., Ng, A.: Reasoning with neural tensor networks for knowledge base completion. In: Burges, C.J.C., Bottou, L., Welling, M., Ghahramani, Z., Weinberger, K.Q. (eds.) Advances in Neural Information Processing Systems 26, pp. 926–934. Curran Associates, Inc. (2013)
10. Wang, Q., Mao, Z., Wang, B., Guo, L.: Knowledge graph embedding: a survey of approaches and applications. IEEE Trans. Knowl. Data Eng. **29**(12), 2724–2743 (2017)

Exploitation of Open Data Repositories for the Creation of Value-Added Services

Antonio Sarasa Cabezuelo[(✉)]

Complutense University of Madrid,
C/Profesor José García Santesmases, 9, 28040 Madrid, Spain
asarasa@ucm.es

Abstract. Open data repositories are information repositories that offer data that can be accessed by anyone using a typical REST type web services API. The API allows sending GET requests to obtain the desired data and return based on the requests made, data sets in various data formats. These repositories can be both public and private. In Spain you can find a large number of public open data sources. For example, the open data source of the City of Madrid offers a large amount of data of different kinds such as traffic accidents, CO_2 emissions, noise pollution, etc. In this article we propose a model for the production of value-added services that did not previously exist using open data. This model is illustrated by creating a service aimed at creating tourist visits in the city of Madrid using open data about museums, temples and monuments found in the open data source of the City of Madrid.

Keywords: Open data repositories · Value-added services · Tourist visits

1 Introduction

This article analyzes the possibilities of reusing and processing information found in the network to build value-added services from them. The Web can be considered the largest repository of information that allows queries to retrieve information. In this sense, the data has become an element of great value due to the possibilities of obtaining information from the processing thereof and to use it for different economic, strategic purposes [1]. A lot of initiatives have promoted the development of actions aimed at making public on the web the information and data they possess public [2]. This article will focus on specific initiative: Open data.

The concept of value-added service [3] refers to applications that are created based on data that come from different sources of information. This data is processed and offered to specific groups of users through a set of access and consultation methods [4]. Some examples of this type of application are hotel search websites, trips, shows… This type of applications have their origin in the so-called Mashup [5]. A key element that has influenced its spread has been the development of web services technology of the APIS-REST type [6]. This technology offers advantages such as a simple information exchange process [7], allows sources to publish the information they want without having to give full access to all available information (access to information is done through web services specific [8] and it is not possible to perform other types of

© Springer Nature Switzerland AG 2020
F. Herrera et al. (Eds.): DCAI 2019, AISC 1003, pp. 134–141, 2020.
https://doi.org/10.1007/978-3-030-23887-2_16

queries), and offer a simple mechanism to invoke services through a simple url with parameters and to retrieve them in the form of a file with a specific data format. This system implements a consumer-producer model [9], which allows creating applications with very different purposes but using the same data from the same information sources, but they differ in the way they process and visualize the information [10].

The open data repositories [11] constitute an initiative arising in the field of public institutions and large companies. Its main objective is to provide anyone with the possibility of the data generated in the activities carried out by these institutions. These data are provided in different electronic data formats such as xml, json, csv and others [12]. To do this, the data is digitized and stored in digital information repositories. Likewise, in order to access the repositories to consult the information, specific web portals that offer information retrieval and retrieval services are created above the repositories. There are several types of access services [13]. Thus, the most basic service consists of a catalog where the different sets of open data that are offered are categorized according to a taxonomy with few levels of depth. In this way, the search is done using a search engine based on keywords or the taxonomy with which the classification is used to implement a faceted search. The result of the search is a file with the requested data represented using one of the commented formats. An example of an open data portal that uses a data catalog is the open data initiative of the Government of Spain [14]. Another type of more advanced service to perform searches is to use a REST API [15] of web services. This mechanism is the most used by technology companies such as Google [16], Twitter (https://developer.twitter.com/en/docs/api-reference-index), Facebook [17] (https://developers.facebook.com/docs/graph-api/using-graph-api) or LinkedIn (https://developer.linkedin.com/docs/rest-api). The APIs act as a query language that is invoked using a set of parameters that configure aspects about how the query will be carried out, the format of the results to be returned, the data sets that will be accessed …

The web service has the same structure as a web address [8]. In this sense it is a url with the peculiarity that at the end of it, the configuration parameters are added in the form of key-value tuples joined by the "&" symbol. In addition, each web service refers to a resource on the server. The resource consists of a script that implements the program that performs information retrieval. Therefore, web services are invoked from a web browser. The result of the invocation is the execution of a script on the server, returning as a result a file containing the recovered data (for example, the invocation of the geocoding service of the Google Maps REST API returns a json file). Finally, note that the portals that use this consultation mechanism offer two access modalities. One modality consists of free access to any web service, and another modality consists of the free execution of certain services under certain conditions and for the rest of services a fee must be paid for each inquiry made.

This article proposed a general process model for the exploiting open data repositories and introduces an application that allows access to the information of Madrid's painting museums, and planning visits. In order to implement, sources of open data about the museums of the city of Madrid are also consulted for this work.

The structure of the paper is as follows. In Sect. 2 is described the general process model. Section 3 introduces the implementation of an example about the general process model. And finally, a set of conclusions and lines of future work will be presented.

2   A Process Model for Development

In this section a proposal is made of a general process model for the creation of value-added services by exploiting open data repositories. The objective is to formalize the workflow that is necessary to implement a sustainable production model of services where the contents are maintained and obtained from data sources that any user can use freely. When you create applications that manage data you spend a lot of time keeping the data: get it, retrieve it and keep it. In this sense, with this proposal, all the problems referred to the data are transferred to the institution that maintains the data, so that the programmer is liberalized from this task. In this sense, the model aims to automate and structure production.

In this model there are 4 key elements that must be taken into account:

- The type of service that you want to create.
- The type of open data sources that is necessary.
- The type of access offered by open data sources.

In this model there are 3 main actors. First, there is the institution that owns the open data sources that are going to be consulted, and that is in charge of offering access to them. In second place is the computer that will be responsible for creating the added value service, which is understood as a software layer that will exploit the data to make them useful for users. In third place is the final user who will be the one who will make use of the data of the repositories in an implicit way through the functionality created by the computer. The scheme in Fig. 1 shows that the model has 7 activities.

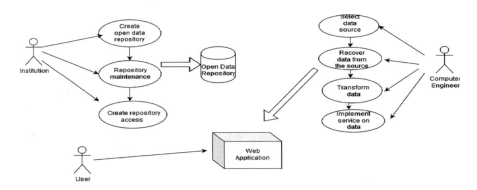

Fig. 1. A process model for development.

3   Mining of Open Data

3.1   Open Data Portal of the Madrid City Council

The city council of Madrid has an open data portal [18] that can be found at the following address: https://datos.madrid.es/portal/site/egob/. The access to the data is done through a catalog where 329 data sets appear. The catalog can be downloaded in

RDF or CSV format. Each sector is shown on its dataset, the date on which it was included in the catalog, its frequency, the number of total downloads and the formats in which it is available. Figure 1 shows an image of the portal.

Fig. 2. The open data portal of the Madrid City Hall.

The search on the portal can be done in two different ways: directly navigating on the catalog or using a filtering system (Fig. 2). In this sense, you can make filters using the elements that are displayed in 'Filter by …' to limit the results to certain sectors, formats and journals, as well as sort by name and date of incorporation. the data to be consulted can be downloaded directly in the form of a file or a REST API can be used that offers a set of consultation services, as shown in Fig. 3a and b is an example of access to a set of data using the two possibilities.

3.2   A Value-Added Service for Tourism in Madrid

From the sources of open data of the city council of Madrid a web application has been developed that implements a value added service that allows to create cultural visits in the context of the city of Madrid. The application has been developed using the Express application framework, and has been modeled as a client/server architecture. The client of the application has been designed using templates ejs. The code used for the ejs templates is HTML5, which is responsible for defining the structures of the page and CSS3 code, which is responsible for giving styles to these structures. In this sense to facilitate the development of the CSS3 code Bootstrap has been included, whose styles are used in practically all the templates, so that the own css is only used for corrections or small modifications of the Bootstrap code.

To facilitate the work of the server, Javascript (together with JQuery and Ajax) has been included in the client, thus freeing the server of a large part of the application load and delegating to the client some functions such as:

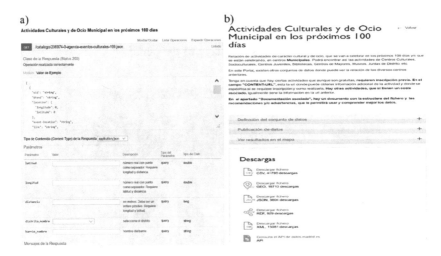

Fig. 3. (a) Access using API -REST, (b) Access using direct download.

- Validation of forms: JavaScript and JQuery are used to validate the forms that appear throughout the application.
- Calls to the server: Ajax is used to make calls to the server in a dynamic and interactive way without needing to reload the client and thus allowing a greater sense of fluidity and speed in the client.
- Dynamic code construction: Much of the design of the application uses this type of coding where the data is obtained through an Ajax call to the server. To avoid doing a reload of the page, the HTML elements are dynamically inserted along with their associated styles to the page using JQuery.
- Calls to APIS: To show some additional information regarding some monuments or visits, calls are made to APIS for external data. In this sense, to avoid overloading the server is the client who performs them directly.

In the Fig. 4 the architecture of the application is shown.

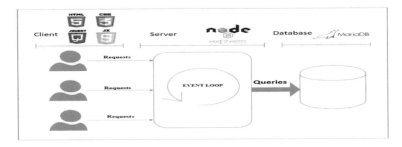

Fig. 4. Application architecture

In the application, three different types of actors are differentiated:

- Visitor: Access the application without having to authenticate and can execute operations that do not require data persistence.
- Registered user: This actor must be registered in the system. Carry out all the actions of the visitor, in addition to their own, including those that need persistence in the data.
- Administrator: This actor must be registered in the system. It carries out maintenance operations of the system, although it can also perform the actions of a visitor.

Figure 5 shows the functions of each of the users.

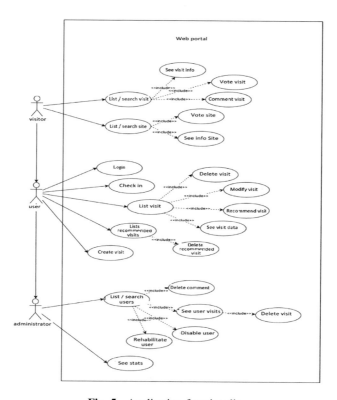

Fig. 5. Application functionality

For example, the functionality of creating a visit can be considered. First the client requests access to create a visit from the portal of the application and a form is shown (Fig. 6a). Once the user has filled in all the fields and has selected the option to create, then the Checks on the client side to ensure that the fields have been properly completed. In case of success, the page will be reloaded showing the visit created (Fig. 6b) and in case of error will be reported.

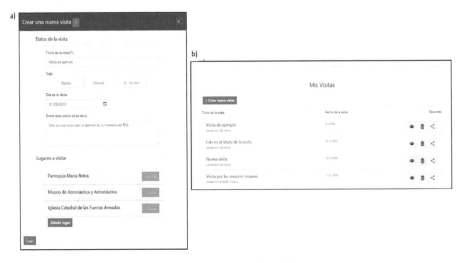

Fig. 6. (a) Creation form, (b) Visit created.

4   Conclusions and Future Work

The use of open data sources provides a useful way to create applications that offer services by simply using the data creatively. In this way you can create an application with an extensive database whose maintenance and updating is done by a third party, thus allowing all the effort to focus on the development itself.

A web application has been implemented using the Node technology to verify the potential and scope of using open data, since without having to enter data in a database, there is access to an immense amount of data on the different tourist monuments of the city of Madrid.

The main lines of future work that could be followed to expand the scope of work:

- Develop the client using a framework such as Angular or React.
- Include additional open data sources that provide more information about the monuments and thus complement the visits.
- Carry out a test evaluation with potential users of the application, to detect possible design flaws or expand its scope based on the conclusions of the evaluation.

Acknowledgment. This work was partially funded by eLITE-CM S2015/HUM-3426, RedR +Human (TIN2014-52010-R) and CetrO+Spec (17-88092-R) projects. I would like to thank this work to Eduardo Vela Galindo.

References

1. Larson, R.R.: Introduction to information retrieval. J. Am. Soc. Inf. Sci. Technol. **61**(4), 852–853 (2010)
2. Lee, Y.J., Kim, J.S.: Automatic web API composition for semantic data mashups. In: 2012 Fourth International Conference on Computational Intelligence and Communication Networks (CICN), pp. 953–957. IEEE (2012)
3. Brunswicker, S., Majchrzak, A., Almirall, E., Tee, R.: Cocreating value from open data: from incentivizing developers to inducing cocreation in open data innovation ecosystems. In: World Scientific Book Chapters, pp. 141–162 (2018)
4. Lehmann, J., Isele, R., Jakob, M., Jentzsch, A., Kontokostas, D., Mendes, P.N., Bizer, C.: DBpedia–a large-scale, multilingual knowledge base extracted from Wikipedia. Semantic Web **6**(2), 167–195 (2015)
5. Gandon, F.: A survey of the first 20 years of research on semantic Web and linked data. Revue des Sciences et Technologies de l'Information-Série ISI: Ingénierie des Systèmes d'Information (2018)
6. Masse, M.: REST API Design Rulebook: Designing Consistent RESTful Web Service Interfaces. O'Reilly Media, Inc., Sebastopol (2011)
7. Gouveia, F., Lira, S.: Semantic approaches for the use of cultural data. Int. J. Comput. Methods Heritage Sci. (IJCMHS) **2**(2), 75–96 (2018)
8. Verborgh, R., Steiner, T., Van Deursen, D., Coppens, S., Vallés, J.G., Van de Walle, R.: Functional descriptions as the bridge between hypermedia APIs and the Semantic Web. In: Proceedings of the Third International Workshop on Restful Design, pp. 33–40. ACM (2012)
9. Vrandečić, D.: Wikidata: a new platform for collaborative data collection. In: Proceedings of the 21st International Conference on World Wide Web, pp. 1063–1064. ACM (2012)
10. Sansonetti, G., Gasparetti, F., Micarelli, A., Cena, F., Gena, C.: Enhancing cultural recommendations through social and linked open data. User Modeling User-Adapted Interact. **29**, 121–159 (2019)
11. Kovalenko, O., Mrabet, Y., Schouten, K., Sejdovic, S.: Linked data in action: personalized museum tours on mobile devices. In: ESWC Developers Workshop, pp. 14–19 (2015)
12. Sato, Y., Kashihara, A., Hasegawa, S., Ota, K., Takaoka, R.: Diagnosis with linked open data for question decomposition in web-based investigative learning. In: Foundations and Trends in Smart Learning, pp. 103–112. Springer, Singapore (2019)
13. Heath, T., Bizer, C.: Linked data: evolving the web into a global data space. Synth. Lect. Semant. Web: Theory Technol. **1**(1), 1–136 (2011)
14. Radulovic, F., Mihindukulasooriya, N., García-Castro, R., Gómez-Pérez, A.: A comprehensive quality model for linked data. Semantic Web **9**(1), 3–24 (2018)
15. Hernández, D., Hogan, A., Krötzsch, M.: Reifying RDF: what works well with wikidata? In: SSWS@ ISWC, vol. 1457, pp. 32–47 (2015)
16. Färber, M., Ell, B., Menne, C., Rettinger, A.: A comparative survey of dbpedia, freebase, opencyc, wikidata, and yago. Semantic Web J. **1**, 1–5 (2015)
17. Bodle, R.: Regimes of sharing: open APIs, interoperability, and Facebook. Inf. Commun. Soc. **14**(3), 320–337 (2011)
18. Shadbolt, N., Gibbins, N.: Resource description framework (RDF). In: Encyclopedia of Library and Information Sciences, pp. 4539–4547 (2010)

Feature Extraction and Classification of Odor Using Attention Based Neural Network

Kohei Fukuyama[1], Kenji Matsui[1(✉)], Sigeru Omatsu[1],
Alberto Rivas[2], and Juan Manuel Corchado[2]

[1] Faculty of Robotics and Design, Osaka Institute of Technology, Osaka, Japan
mlm18r23@st.oit.ac.jp,
{kenji.matsui,shigeru.omatsu}@oit.ac.jp
[2] BISITE Digital Innovation Hub, University of Salamanca, Salamanca, Spain
{rivis,corchado}@usal.es

Abstract. In this paper, we explored a new approach of odor identification using deep learning. We also verified whether it is possible to classify odors in real time. Attention recurrent neural network (ARNN) is mainly used in the sequence to sequence model in terms of natural language processing field. We applied the simple ARNN embedding part with auto encoder to the deep learning model. This model can visualize the attention part of input data. Six types of aroma oil were measured using five types of metal oxide semiconductor gas sensors. Using the measured data, the accuracy was 96.83 [%] for training data and 96.88 [%] for the verification data.

Keywords: Deep learning · Sequence to sequence · Attention mechanism · Odor classification · Visualization · Metal oxide gas sensor

1 Introduction

Olfactory perception studies have been relatively behind among the five sense information processing. Sense of sight, hearing, and touch are looking at physical values such as light, sound, and pressure. Those values are easy to measure, however, in case of odor, it is difficult to observe the dynamic change of chemical values which olfaction research requires. Most of the time, the odor gas concentration is extremely low, and that makes our odor analysis very difficult.

Axel and Buck of Columbia University in 1991 revealed that about 1000 kinds of receptors are present in the olfactory cells [1]. Since then, researchers have been figuring out the fundamental mechanism of olfaction, and the related applied research themes such as detection or identification of odor have been increasing. There are already some products available although the applications are limited.

Omatsu et al. [2] applied LVQ algorithm for odor classification. The result showed 96% accuracy in case of four types of tea, and 89% in case of five types of coffee. In that research, they are looking at the sensor output for both the initial stage vale and the peak value. The LVQ uses the difference of those values.

With the advent of deep learning, research and development on artificial intelligence technology has been actively conducted. Especially, accuracy of image recognition and speech recognition rate has been drastically improved. In case of odor classification,

F. Herrera et al. (Eds.): DCAI 2019, AISC 1003, pp. 142–149, 2020.
https://doi.org/10.1007/978-3-030-23887-2_17

image recognition related approach has been mainly applied. Peng, Pai, et al. [3] applied Deep Convolutional Neural Networks for odor classification. Using four types of gas, the concentration of each gas is classified into 20 stages with an accuracy of 96%.

However, in those methods, it takes a long time to classify because the methods require long time sequence data. If we can classify odors in real time, we will be able to develop many useful applications. For that reason, real time odor classification is very attractive research target.

In this paper, our challenge is to solve the classification processing time problem by applying the method used in the natural language processing field and focusing on the time change. We report our experimental study using six types of aroma oil with five metal oxide semiconductor gas sensors, and visualize, then distinguish feature quantity using the proposed method.

In addition, we verified whether it is possible to classify the odor, if we have 5 s of continuous odor sensor output data, no matter when we obtain the data out of long time sequence odor data.

2 Key Components Used for the Proposed Method

2.1 GRU (Gated Recurrent Unit)

Commonly used LSTM (Long short term memory) can continue using short-term memory for a long time, however, it is high in calculation cost [4]. On the other hand, GRU is a kind of gated RNN, which can alleviate gradient loss at learning. Since our model does not require long-term memory, we are using calculation cost effective GRU. (1)–(4) show how to calculate GRU, and Fig. 1 shows the GRU algorithm.

$$z_t = \sigma(W_z X_t + U_z h_{t-1} + b_z) \tag{1}$$

$$r_t = \sigma(W_r X_t + U_r h_{t-1} + b_r) \tag{2}$$

$$g_t = \tanh\left(W(r_t \odot h_{t-1}) + W_z X_t + b_g\right) \tag{3}$$

$$h_t = (1 - z_t) \odot h_{t-1} + z_t \odot g_t \tag{4}$$

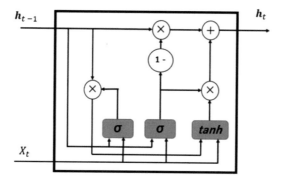

Fig. 1. GRU algorithm

2.2   Seq2Seq Model

The Seq2Seq (Sequence to Sequence) model is also called Encoder-Decoder model. It is a model specialized for time-series data with the time-dependent relationship, mainly used in the natural language processing field. Seq2Seq converts time series data into another time series data. On the encoder side, the input word sequence is converted into One-Hot vector in order, then converted into an intermediate vector. On the decoder side, first input <eos> as a dummy word. For the input of the second step, input the output result of the previous step. As a result, time-series data can be converted. As an example, Neural machine translation is shown in Fig. 2.

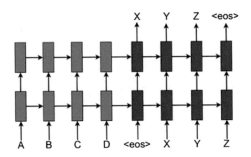

Fig. 2. Neural machine translation [5]

2.3   Attention Mechanism

The attention mechanism is a method to further improve the accuracy of Seq2Seq. By applying attention mechanism, we can pay attention to necessary information just like humans. By calculating the a_t, alignment associated with the output of each step using the intermediate vector $\tilde{h}_s$[5], we can generate the attention map. The method of calculating the output C_t of the attention model and the attention model are shown in Fig. 3.

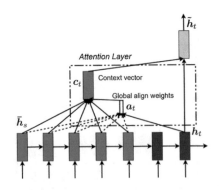

Fig. 3. Global attention model [5]

$$S = \tilde{\boldsymbol{h}}_s \cdot \boldsymbol{h}_t \tag{5}$$

$$\boldsymbol{a}_t = softmax(\boldsymbol{S}) \tag{6}$$

$$\boldsymbol{C}_t = \boldsymbol{h}_s \cdot \boldsymbol{a}_t \tag{7}$$

3   Measurement of Odor Data

We have measured six types of odors as shown in Table 1. Table 2 shows the sensors used for the measurement. The measurement of aromatic oil was carried out as follows.

First, the impure gas in the sampling box in Fig. 4 needs to be exhausted. Since the output voltage of each sensor fluctuates for a while, we need to wait until the voltage becomes stable. The alumina plate temperature is set to 100 °C using the equipment in Fig. 5. The aromatic oil samples are placed 10 μL on the alumina substrate and the sensor outputs are measured until those outputs reach the voltage before the reaction.

Table 1. Aromatic oil

Aroma number	Aroma name	Category
0153	Ocean	Marine
1105	Green Tea & Lemongrass	Floral
1524	Orange Spice	Citrus
1551	White Cashmere	Oriental
1595	Summer Symphony	Citrus
1970	Amalfi Coast	Fine Fragrance

Table 2. Sensor list

Sensor number	Type	Gas detect types
S1	TGS 822	Organic solvent
S2	TGS 825	Hydrogen sulfide
S3	TGS 826	Ammonia
S4	TGS 2602	Ammonia Hydrogen sulfide
S5	TGS 2620	Alcohol organic solvent

Fig. 4. Sampling box

Fig. 5. Alumina substrate

4   Proposed Model

In the proposed model, Auto Encoder was applied to the embedding part of Seq2Seq to extract feature data of the input data. We also added an attention node and visualized which part of the input data we are interested in. We used GRU in the RNN part instead of using LSTM because of the advantage of calculation cost. The encoder side of Seq2Seq is total 5 layers using bi-directional RNN. As for the decoder side, there are 4 layers of GRU, followed by 3 layers of complete connection layer and soft Max layer. There are total 8 layers. Figure 6 shows the proposed model.

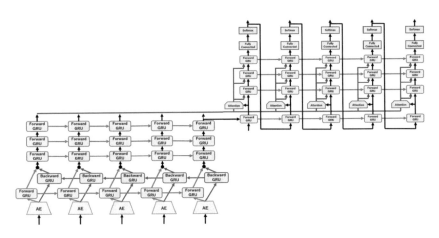

Fig. 6. proposed model

5   Identification Method

As for the Seq2Seq model input data, 5 ms long sensor signals were pre-processed, then applied to the encoder input. Each auto encoder at the input converts five sensor values into three dimensional features. As for the inputs to the decoder part, the identification results are applied sequentially. Attention node focuses on specific parts of the input sequence. The identification result comes from the classification output at the time 5 s past from the initial stage.

6   Result

Using the proposed model, seven condition (six types of aroma oil + non odor) were classified. The accuracy was 96.83 [%] for training data and 96.89 [%] for testing data. Table 3 shows the details of the classification results.

Table 3. Classification results

Classification results (96.89)%									
Odor number	Non odor	No. 0153	No. 1105	No. 1524	No. 1551	No. 1595	No. 1970	Total	Classification rates
Non odor	7889	452	101	200	146	15	179	8982	87.83%
No. 0153	89	12146	8	0	220	0	0	12463	97.46%
No. 1105	201	0	13217	0	0	27	0	13445	98.30%
No. 1524	445	0	0	18330	0	0	0	18775	97.63%
No. 1551	609	21	0	0	16920	15	0	17565	96.33%
No. 1595	381	0	48	0	0	21773	0	22202	98.07%
No. 1970	326	255	12	0	0	0	26366	26959	97.80%

The feature value of the testing data obtained from the Embedding section is as shown in Fig. 7.

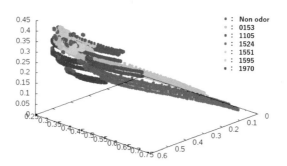

Fig. 7. Feature values obtained from the Embedding section

From Fig. 7, it can be seen that the characteristic quantities of the odor of each aromatic oil are extracted.

Next, the following results were obtained from the Attention section. Figures 8, 9, 10, 11, 12, 13 and 14 show the visualized attention part of the input data by the attention node. Those white parts are focusing section by ARNN at each step.

0, 0, 0, 0, 0, (True, True, True, True, True)

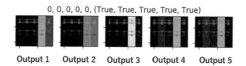

Output 1   Output 2   Output 3   Output 4   Output 5

Fig. 8. The part that is paying attention No smell

1, 1, 1, 1, 1, (True, True, True, True, True)

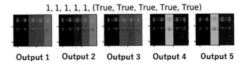

Output 1   Output 2   Output 3   Output 4   Output 5

Fig. 9. The part that is paying attention 0153

2, 2, 2, 2, 2, (True, True, True, True, True)

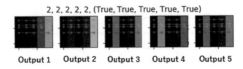

Output 1   Output 2   Output 3   Output 4   Output 5

Fig. 10. The part that is paying attention 1105

3, 3, 3, 3, 3, (True, True, True, True, True)

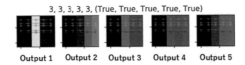

Output 1   Output 2   Output 3   Output 4   Output 5

Fig. 11. The part that is paying attention 1524

4, 4, 4, 4, 4, (True, True, True, True, True)

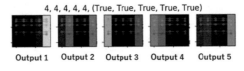

Output 1   Output 2   Output 3   Output 4   Output 5

Fig. 12. The part that is paying attention 1551

5, 5, 5, 5, 5, (True, True, True, True, True)

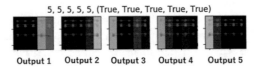

Output 1   Output 2   Output 3   Output 4   Output 5

Fig. 13. The part that is paying attention 1595

6, 6, 6, 6, 6, (True, True, True, True, True)

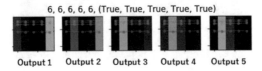

Output 1   Output 2   Output 3   Output 4   Output 5

Fig. 14. The part that is paying attention 1970

From Figs. 8, 9, 10, 11, 12, 13 and 14, you can visualize the parts of attention, depending on the state of each odor.

7 Discussion

As shown in Table 3, the classification result of each six aromatic oil shows a high percentage of correct answers of 96% or more. Those results were obtained from any 5 s segments extracted from the sensor output data. The possible reason of such excellent performance is the introduction of ARNN. However, in the case of non-odor, the correct answer rate dropped to 86%. This is because the sensor output during the transition time shows similar values as in the case of non-odor. In addition, the steady state changes depending on the measurement time and air temperature. These factors are also considered to be able to obtain stable and accurate results. We plan to reduce these problems using preprocessing of input data.

8 Conclusions

In the proposed model, 96% accuracy was obtained. Such accuracy was obtained from any 5 s segments extracted from the sensor output data possibly because of the introduction of ARNN. Also the model automatically extract the feature values of the input data and visualize the attention part so that the relationship of each odor features can be observed clearly. Because the real time classification performance is quite powerful feature, we plan to find some useful applications as well as improving the algorithm.

References

1. Axel, R., Back, L.B.: The Nobel Prize in Physiology or Medicine (2004)
2. https://www.nobelprize.org/nobel_prizes/medicine/laureates/2004/press.html
3. Omatu, S., Yano, M.: E-nose system by using neural networks. Neurocomputing **172**, 394–398 (2016)
4. Peng, P., et al.: Gas classification using deep convolutional neural networks. Sensors **18**(1), 157 (2018)
5. Chung, J., et al.: Empirical evaluation of gated recurrent neural networks on sequence modeling. arXiv preprint arXiv:1412.3555 (2014)
6. Luong, M.-T., Pham, H., Manning, C.D.: Effective approaches to attention-based neural machine translation. arXiv preprint arXiv:1508.04025 (2015)

A Conceptual Group Decision Support System for Current Times: Dispersed Group Decision-Making

João Carneiro[1]([⊠]) [iD], Patrícia Alves[1] [iD], Goreti Marreiros[1] [iD],
and Paulo Novais[2] [iD]

[1] GECAD – Research Group on Intelligent Engineering and Computing
for Advanced Innovation and Development, Institute of Engineering,
Polytechnic of Porto, 4200-072 Porto, Portugal
{jrc,prjaa,mgt}@isep.ipp.pt
[2] ALGORITMI Centre, University of Minho, 4800-058 Guimarães, Portugal
pjon@di.uminho.pt

Abstract. We are living a change of paradigm regarding decision-making. On the one hand, there is a growing need to make decisions in group at both professional and personal levels, on the other hand, it is increasingly difficult for decision-makers to meet at the same place and at the same time. The Web-based Group Decision Support Systems intend to overcome this limitation, allowing decision-makers to contribute to the decision process anytime and anywhere. However, they have been defined inadequately which has been compromising its success. In this work, we propose a conceptual definition of a Web-based Group Decision Support System that intends to overcome the existing limitations and help them to affirm as a reliable and useful tool. In addition, we address some crucial topics, such as communication and perception, that are essential and sometimes forgotten in the support of dispersed decision-makers. We concluded that there are still some limitations, mostly in terms of models and applications, that prevent the design of higher quality systems.

Keywords: Group Decision Support Systems ·
Dispersed group decision-making · Dialogue games · Affective computing

1 Introduction

A group decision-making process consists in a process in which a group of people act collectively in order to select one or more alternatives to solve a certain problem [1]. At first sight, it seems to be an easy process. However, the reality tells us the opposite. Firstly, there are a set of possible alternatives which are defined according to a set of, usually conflicting, criteria. Secondly, there are too many psychological factors that affect the process, such as: social comparison, different intentions, interpersonal relationships, among others [2]. Lastly, if we consider the existence of all these issues in a dispersed context the complexity increases exponentially [3].

The Web-based Group Decision Support Systems (GDSS) have been studied since the beginning of the 21st century and they intend to support the group decision-making

© Springer Nature Switzerland AG 2020
F. Herrera et al. (Eds.): DCAI 2019, AISC 1003, pp. 150–159, 2020.
https://doi.org/10.1007/978-3-030-23887-2_18

process anytime and anywhere [4]. They distinguished from conventional GDSS because they operate on the Web, which make them available by simply having an internet connection [5, 6]. However, if the general opinion is that these systems are crucial to the current times, the fact is that they have been struggling to impose [4, 7]. The research under this topic has been mostly oriented to study models that are capable of proposing solutions according to the decision-makers' preferences. Still, a group decision-making process is much more than just an outcome [8]. In face-to-face meetings, decision-makers communicate (through verbal and nonverbal communication) in order to exchange perspectives, allowing them to reason, to argue, and to create new intelligence [9]. It is all this interaction that composes the process that makes face-to-face meetings advantageous when compared with individual decision-making [10]. Therefore, a system that does not allow decision-makers to benefit from these advantages, will not be seen as a valuable asset and consequently as something worth to use. Even if just hypothetically, it is capable of proposing the best solution for a certain problem according to the decision-makers' preferences, but it is not capable of "explaining" the reasons behind that proposal, two things can happen: (1) the system will not be seen as reliable and the proposal can be seen as some kind of guess and (2) this behavior would impede the creation of new intelligence which annihilates all the advantages associated to conventional group decision-making.

The aim of this paper is to present a conceptual Web-based GDSS especially designed for dispersed group decision-making. The proposed approach presents a set of essential features to achieve the system's success and acceptability. Considering these features, we also propose a set of strategies to implement them. In addition, some important topics that are sometimes ignored under the group decision-making context are addressed.

The rest of the paper is organized in the following order: in the next Section, we describe the proposed Web-based Group Decision Support System's conceptual model, mostly in terms of features and architecture. In Sect. 3, we present a conceptual strategy to represent decision-makers under dispersed contexts. In Sect. 4, we analyze possible perspectives to support and recommend dispersed decision-makers in the decision-making process. Finally, some conclusions are put forward in Sect. 5, alongside with suggestions of work to be done hereafter.

2 The Web-Based Group Decision Support System's Conceptual Model

This work distinguishes essentially by the way how the problematic of group decision-making support is addressed. Rather than idealizing the system in turn of a model capable of proposing a solution according to certain configurations, our focus is on allowing decision-makers to benefit from the typical advantages associated with face-to-face group decision-making processes.

We propose a Web-based GDSS inspired in the behavior of a social network like Facebook or LinkedIn. That means, the system should potentiate the interaction between decision-makers. The issues are discussed in the form of topics and everyone can add comments and replies. The system's main features are (Fig. 1): the ability to foster communication between decision-makers; the ability to represent decision-makers in terms of preferences, intentions, goals, desires, interests, beliefs, social standing, credibility and expertise; to help decision-makers perceive the process in terms of how the decision and everything that composes the context evolves over time; and finally, a set of strategies that support decision-makers through proposals, recommendations, predictions, relevant information, among others.

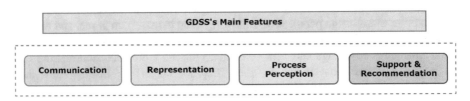

Fig. 1. Main features of the proposed Web-based GDSS.

2.1   Conceptual Flow

Communication is the key ingredient of a group decision-making process. Thus, we propose a system that potentiates the communication between decision-makers. Obviously, it is different to communicate in face-to-face contexts and through an online application [11, 12]. So, the communication should be more structured than the one practiced in presential contexts. With a more structured type of communication the system will be able to make use of the conversations made by decision-makers to support the decision-making process and for autocomplete purposes, for instance, in the definition of multi-criteria problems or/and in the identification of alternatives and criteria. Another important aspect is that internet users are already used to social networks, which facilitates the understanding that each subject should be debated on a different topic. Other important strategies such as the use of "Likes (thumbs up)" and other forms of expression can also be used, since it is something that people are already used to and can serve as strategies that allow to better understand the level of acceptance of different ideas and the level of importance of the different subjects. Figure 2 represents (in a non-formal format) the activities carried out by decision-makers in the use of our conceptual proposal. As we can see, decision-makers can communicate even if they are not (yet) involved in a decision-making process. We consider that the identification of a problem can occurs in a normal dialogue. So, this is the first step to start a decision-making process. After that, a decision-maker can create a new problem or submit a ticket asking a facilitator to create the group decision-making (GDM) problem. When a problem is created and the participants are added to the process, each participant can then start by some initial configurations. This is the first time that decision-makers (in general) interact with the new decision-making process and where they can define important stuff such as: their expertise level, their intentions and point other decision-makers as experts in that topic. In this way, they are modelling their representation and helping the system understanding all the context. All these steps can be revisited at any time and decision-makers can perform many reconfigurations as they want. In fact, the system can make use of these changes to better understand the process and consequently to create intelligence. After that, and if alternatives and relevant criteria are not yet defined, the system should provide conditions to perform the idea generation step. The most structured type of communication used by the means of the system will help in the organization of different ideas, in the identification of alternatives and of the most important criteria.

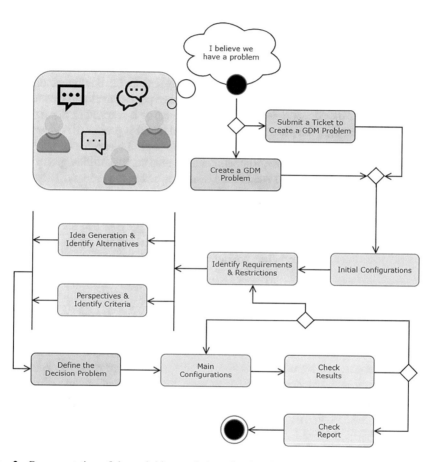

Fig. 2. Representation of the activities carried out by decision-makers in the use of the system.

After all the different possible alternatives are identified, the decision problem should be defined, for that Artificial Intelligence techniques can make use of the dialogues performed by decision-makers in the previous steps. Techniques to extract knowledge can be used in here. The definition of the decision problem should not be costly because the different alternatives were in majority referred before, such as the most important criteria. When the decision problem is defined, decision-makers can then configure their preferences regarding not only alternatives and criteria, but also regarding limitations, desires, goals and their group position in terms of opinion. This seems complex but it can be done through simple "clicks" using configuration templates with high usability. Right after decision-makers perform their configurations, many Artificial Intelligence techniques can be used to propose solutions, to search inconsistencies, to present relevant information and to support/recommend decision-makers. Decision-makers do not need to be aware of this level of complexity, but the strategy used to present the information to them is vital. That means, the communication between the system and each decision-maker should be adapted to his/her preferences and interests, and the system should be capable of understanding how much each decision-maker is involved in the process. Obviously that the capacity to propose solutions is intrinsic to this kind of systems, but in this conceptual proposal the major objective is the walk to find the solution. The system should be aware of the process importance, which means that the system should be intelligent enough to understand how important to mature ideas, to exchange perspectives and to reflect is, in order to potentiate the decision quality. When a consensus is reached or the satisfaction level attained is enough, the process ends, and a final report is presented. Otherwise, the previous steps can be revisited. Finally, the system should be capable of using all the data generated during the process, to document the reasons that led the group to make that decision and the impact that each decision-maker had in the process. This will turn into valuable information for the organizations because they can, in the future, understand the reasons that made them take those decisions and the responsibility/contribution that each decision-maker had (either for the good or bad decisions). Also, Artificial Intelligence techniques can be used in these reports to learn from past experiences in order to make better predictions and recommendations.

2.2 Architecture

The literature is not rich in terms of architectures for Web-based GDSS, though in a first instance, a Web-based GDSS differs from a conventional GDSS mostly because of its architecture. In this work, we propose a microservices-based architecture (Fig. 3) because it empowers a lot of benefits for the context of the proposed system (group decision-making with dispersed participants) and for the current context of the major organizations. If we think in terms of the number of features a system like this has to provide (and the number of different algorithms and models used), a microservices-based architecture allows to: get a better faults isolation, perform continuous delivery, have components spread across multiple servers, be easily understood since they represent small pieces of functionality, etc.; and from the organizations perspective it allows to: organize the code around business capabilities, use complement cloud activities, write code in different languages, get an easy integration, perform automatic deployment, etc. [13].

Figure 3 represents the conceptual Web-based GDSS proposed in this work. It uses a standard microservices architecture. There is an API Gateway that works as a single-entry point into the system, which allows the internal system architecture to be encapsulated and to provide an API that tailors to each client. In addition, we consider that features such as authentication, monitoring, load balancing, among others, are also responsibilities of the API Gateway.

We consider the existence of a set of possible microservices/services that satisfies the business of the organization (accounts, products, etc.) and a set of microservices that intends to support the decision. In this conceptual proposal we consider several Artificial Intelligence's strategies (which are explored in Sect. 4). Each strategy has special needs and is implemented using different programming languages. Thereby, a microservices-based architecture becomes even more relevant because this way each service is independently deployable, loosely coupled, highly maintainable and testable, easier to understand and is relatively small.

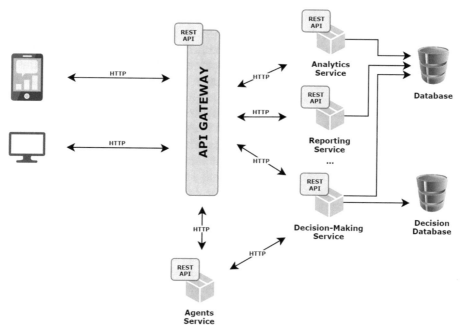

Fig. 3. The Group Decision Support System's conceptual architecture.

In this proposal, we pay particular attention to the "Agents Service" and to the "Decision-Making Service". The "Agents Service" is the microservice that encapsulates the Multi-Agent System existent in the Web-based GDSS and where the main agents' platform stands. In the "Agents Service" the information that circulates in the system is analyzed and processed by the agents when necessary. The "Decision-Making Service" is responsible for the strategies used to automatically propose solutions and where exists another agents' platform to work with contextual agents (or clones like we will see in Sect. 3).

3   Decision-Makers Representation

In this conceptual approach, we consider some aspects regarding the decision-makers representation. We propose the use of a multi-agent system in which each decision-maker is represented by an agent. From now on we call each of these agents as participant agents. As has been said, a participant agent represents a decision-maker, that means a participant agent should be modelled with the characteristics of that decision-maker in order to represent him/her accordingly. However, there are three aspects that affect exponentially the complexity of modeling agents in this context. First, decision-makers behave differently according to the situation's characteristics (for instance, how much a decision problem means to them) [14], second, these different behaviors are always affected by the decision-makers' personality traits (which are more constant) and third, the decision-makers' knowledge evolves over time.

In Sect. 2.1 we referred that the multi-agent system is part of the "Agents Service", however, this is not the only agents' platform in the proposed Web-based GDSS. We consider the existence of another agents' platform in the "Decision-Making Service". So, the participant agents in the "Agents Service" represent decision-makers in terms of what they are, more specifically, these agents are modelled with personality traits of the decision-makers they represent and each one has a knowledge base, where the history of everything that matters is saved. It is considered that the participant agents in the "Agents Service" are always active and can be working even when the decision-makers they represent are not involved in any process. They can be processing the decision-makers dialogues, for instance, to study relationships, the evaluations made by other decision-makers to their comments, who usually support them, who usually criticized them, among others. Regarding the agents in the "Decision-Making Service", they represent decision-makers in a context, and they are alive only during the existence of that process. These agents can be modelled with other characteristics, such as: the decision-maker's intentions, which other decision-makers they consider credible, decision problem preferences, etc. These agents can use other Artificial Intelligence techniques in order to better understand the process impact, such as an emotional model. In terms of flow, when a decision-maker is added to a new decision problem, a new participant agent is created in the "Decision-Making Service". This participant agent is a clone of the participant agent existent in the "Agents Service". This approach presents several benefits because it avoids problems when a decision-maker is involved in too many decision problems at the same time, it separates the impact that each process has in the participant agent and makes it easier to manipulate the temporary knowledge and the knowledge that should be persistent.

4   Group Support and Recommendation: Possible Perspectives

Support and recommendation go hand in hand in terms of systems being capable of helping users attaining something better. Decision support techniques and recommendation techniques can both be used to help decision-makers during the process. In this proposal, we consider a set of different techniques to support decision-makers in the group decision-making process. However, due to the nature of the proposed architecture, the system is prepared to grow up and new strategies can be included over time. Moreover, as the proposed architecture is microservices-based, it becomes a lot easier, in terms of scalability and reusability, to work with containers and to monitor them.

As we have seen, a Web-based GDSS should provide good enough conditions so decision-makers can communicate. In this conceptual proposal, this is accomplished through the natural interaction existent in a social network.

Besides communication, there are other tasks that the system should be able to facilitate, such as the creation of new problems. It is known that to define a new decision problem is complex and time-consuming. There are always several alternatives, different criteria types, several decision-makers, etc. So, it is important to use at least 2 different strategies for this purpose: (1) text analytics, to automatically suggest based on the previous dialogues, alternatives, criteria and decision-makers, and (2) algorithms such as case-based reasoning, to predict the possible alternatives and criteria types based on previous problems.

Regarding the strategies to automatically propose a solution to the group, we consider in a first instance a dialogue-based argumentation model and in a second instance a multi-criteria decision analysis (MCDA) model. The former intends to be capable of proposing solutions and at the same time, due to its self-nature characteristics, to be capable of explaining the reasons that lead to the proposition of those solutions. This allows decision-makers to feel part of the process and to understand it accordingly. As they understand the motifs behind the proposed solutions, they can reason about those motifs and consequently it becomes easier to accept or to reject the proposed solutions. We consider a dialogue-based argumentation model with a high level of expressiveness, that means, the participant agents are capable of behaving according to different intentions in the same dialogue. In addition, the participant agents can use the same locution for different purposes (for instance, to persuade or to deliberate) and the identification of these intentions is a responsibility of the other participant agents. The latter intends to work as a detector of inconsistencies. It is extremely important to perceive if the preferences configuration made by each decision-maker makes sense. For instance, let us imagine a decision-maker that considers the price as the most important criterion, but at the same time his/her preferred alternative is the most expensive; it is important to understand if there are not big price differences between alternatives or if he/she made a mistake in the configuration process or if there are subjective reasons behind his/her configuration. We consider that a MCDA model can be extremely helpful in the detection of these inconsistencies.

Considering the number of messages exchanged in a system like this, it is fundamental to study the produced dialogues. For that, we consider the use of two main strategies: text analytics and natural language processing. In this way, the system can study the dialogues and produce important information regarding not only the dialogues structure but mostly in terms of their meaning and the sentiment existent in them. In addition, classic algorithms from the social networks' literature can be used to understand the impact/importance of each message/topic. This information can be used by each agent in the dialogues to better represent their decision-makers' needs/intentions.

Finally, it is important to use strategies that allow to learn, classify, predict and recognize patterns. For that, we consider the use of machine learning algorithms, more specifically, deep learning and reinforcement learning algorithms. With these algorithms each participant agent (in the "Agents Service") will be capable of presenting important information about other decision-makers, about the decision processes (previous and actual) and about things that matter in general to the decision-maker it represents.

5 Conclusions and Future Work

Web-based GDSS have been studied in the last years in order to develop solutions capable of supporting decision-makers anytime and anywhere. They differentiate from conventional GDSS because they operate on the web and so, decision-makers only need to have an internet connection to use them. However, GDSS in general are having problems to establish and to be recognized as a useful tool by organizations. We believe that this is due to the way they have been addressed, which is focused in the outcomes rather than in the decision-makers' needs.

In this paper we propose a conceptual Web-based GDSS that intends to enable decision-makers to benefit from the typical advantages associated to face-to-face decision-making. Our approach allows decision-makers to interact as people do in social networks, which naturally promotes the communication and the interaction between them. In addition, our proposal is based in a microservices architecture that demonstrates a lot of benefits in a system where so many different Artificial Intelligence techniques are implemented. However, we found some limitations that impede to implement an approach like this now, such as: rudimentary tools to develop multi-agent systems, lack of psychological models that can be computerized, and dialogue-based argumentation models not intelligent enough.

As future work, we want to continue digging under the topic of GDSS. After identifying a set of issues that are compromising the GDSS success and a set of possible solutions, we now intend to work in order to make this conceptual model real. We intend to develop dialogue-based argumentation models in which not only the messages exchanged by the agents are perceptible to decision-makers but also the intentions behind the messages.

Acknowledgements. This work was supported by the GrouPlanner Project (POCI-01-0145-FEDER-29178) and by National Funds through the FCT – Fundação para a Ciência e a Tecnologia (Portuguese Foundation for Science and Technology) within the Projects UID/CEC/00319/2019 and UID/EEA/00760/2019.

References

1. Liu, W., Dong, Y., Chiclana, F., Cabrerizo, F.J., Herrera-Viedma, E.: Group decision-making based on heterogeneous preference relations with self-confidence. Fuzzy Optim. Decis. Making **16**, 429–447 (2017)
2. Lerner, J.S., Li, Y., Valdesolo, P., Kassam, K.S.: Emotion and decision making. Ann. Rev. Psychol. **66**, 799–823 (2015)
3. Forsyth, D.R.: Group Dynamics. Cengage Learning, Boston (2018)
4. van Hillegersberg, J., Koenen, S.: Adoption of web-based group decision support systems: experiences from the field and future developments. Int. J. Inf. Syst. Project Manag. **4**, 49–64 (2016)
5. Desanctis, G., Gallupe, R.B.: A foundation for the study of group decision support systems. Manag. Sci. **33**, 589–609 (1987)
6. Gray, P.: Group decision support systems. Decis. Support Syst. **3**, 233–242 (1987)

7. van Hillegersberg, J., Koenen, S.: Adoption of web-based group decision support systems: conditions for growth. Procedia Technol. **16**, 675–683 (2014)
8. Carneiro, J., Saraiva, P., Conceição, L., Santos, R., Marreiros, G., Novais, P.: Predicting satisfaction: perceived decision quality by decision-makers in web-based group decision support systems. Neurocomputing **338**, 399–417 (2019)
9. Dennis, A.R.: Information exchange and use in small group decision making. Small Group Res. **27**, 532–550 (1996)
10. Michaelsen, L.K., Watson, W.E., Black, R.H.: A realistic test of individual versus group consensus decision making. J. Appl. Psychol. **74**, 834 (1989)
11. Carneiro, J., Martinho, D., Marreiros, G., Jimenez, A., Novais, P.: Dynamic argumentation in UbiGDSS. Knowl. Inf. Syst. **55**, 633–669 (2018)
12. Carneiro, J., Martinho, D., Marreiros, G., Novais, P.: Arguing with behavior influence: a model for web-based group decision support systems. Int. J. Inf. Technol. Decis. Making (IJITDM) **18**, 517–553 (2019)
13. Villamizar, M., Garcés, O., Castro, H., Verano, M., Salamanca, L., Casallas, R., Gil, S.: Evaluating the monolithic and the microservice architecture pattern to deploy web applications in the cloud. In: 2015 10th Computing Colombian Conference (10CCC), pp. 583–590. IEEE, (2015)
14. Carneiro, J., Saraiva, P., Martinho, D., Marreiros, G., Novais, P.: Representing decision-makers using styles of behavior: an approach designed for group decision support systems. Cogn. Syst. Res. **47**, 109–132 (2018)

Distributed Computing, Grid Computing, Cloud Computing

Density Classification Based on Agents Under Majority Rule: Connectivity Influence on Performance

Willyan Daniel Abilhoa[2(✉)] and Pedro Paulo Balbi de Oliveira[1,2]

[1] Faculdade de Computação e Informática,
Universidade Presbiteriana Mackenzie, Rua da Consolação 896,
Consolação, São Paulo, SP 01302-907, Brazil
[2] Pós-graduação em Engenharia Elétrica e Computação,
Universidade Presbiteriana Mackenzie, Rua da Consolação 896,
Consolação, São Paulo, SP 01302-907, Brazil
abilhoa.willyan@gmail.com, pedrob@mackenzie.br

Abstract. The density classification task is a prototypical consensus problem of distributed solution, usually addressed in the field of cellular automata. In short, this problem consists of finding the most frequent state in a binary sequence, necessarily through a non-global process on which the automaton reaches uniform consensus about such state. In this regard, we formulate the task as an agent-based model, in which agents set up a connectivity pattern, here corresponding to a circulant graph, and update their internal states according to the majority rule. The performance of the model corresponds to the number of correctly classified densities, given a set of binary sequences. Therefore, our goal is to analyze the sensibility of the model's performance in terms of the connectivity pattern associated with it, configured as a circulant graph, under different orders, average degrees and connectivity arrangements.

Keywords: Distributed problem solving · Emergent computation ·
Density classification · Majority rule · Agent-based models · Circulant graphs

1 Introduction

Distributed problem solving (DPS) [6] is a computing methodology for solving problems that require the collective effort of several integrant units, such as agents, which can interact and perform joint actions without any central control, so that an answer is obtained as a result of these actions. If not for the collective work, a single unit might not have the ability to provide the solution all by itself, or maybe not as fast as necessary, due to the lack of information about the overall situation of the model towards the solution.

A*gent-based models* (ABM) [2] consist of describing a system from the perspective of its constituent units, the agents. In this sense, the interest is on discovering emergent behavior, that is, a process outcome difficult to predict or understand by simply looking at the units individually.

By these two definitions above, it is comprehensible that many ABMs have been applied to problems to be solved by distributed means, where the corresponding

F. Herrera et al. (Eds.): DCAI 2019, AISC 1003, pp. 163–170, 2020.
https://doi.org/10.1007/978-3-030-23887-2_19

solution is only achieved by a process based on the coordinated strategy of these agents and their local and decentralized actions, even if not necessarily simple [2].

The *density classification task* (DCT) [5, 15] is an example of problem to be addressed in this scenario. DCT is the prototypical computational decision problem in which, given a randomly generated sequence of binary states, it is required to answer which state is more prevalent in the sequence, such that this answer has to be necessarily achieved as a result of collective local computations.

Even though DCT is a prototypical problem, efforts to tackle it are quite a challenging issue, as the task is impossible to be solved in the standard formulation model [11]. In this sense, many proposals in literature try to find the best possible performing imperfect solution in the standard formulation (as in [20]), or devising alternative formulations where a perfect solution does exist, as for instance [9]. A wide review of these works is available in [5].

In this regard, we formulate the DCT as an ABM that updates the states of its agents according to a simple *decision-making rule* – the *majority rule* – and their *connectivity pattern*, defined by a *circulant graph* within a *connectivity space*. By these changes, they seek to set up the best possible communication channel for sharing their local information (their local density), in order to accomplish the task, thus determining the global density.

Therefore, the goal of the present work is to determine how the different configurations of circulant graphs, assumed as the connectivity pattern of the agents' interactions, may affect the performance in solving the DCT.

The paper is organized as follows. Section 2 gives a brief contextualization about density classification task, majority rule and circulant graphs. Section 3 explains the formulation as an agent-based model, and brings to context what is necessary in order to classify density of a binary sequence. Section 4 covers the central topics taken into account for the analysis of performance sensitivity to connectivity, as it discusses on the obtained results. Section 5 brings the overall conclusion about the analysis as well as some insights are provided on further possible works related to the results of our experiments.

2   A Brief Contextualization

2.1   Density Classification Task

DCT refers to finding the most frequent state in a cyclic sequence of n binary states $C_0 = \{s_1, \ldots, s_n\}$, interpreted as the *initial configuration* (IC), n being an odd number, usually binary, by means of a non-global procedure [4, 5, 14]. From that, the goal of the task is to determine whether the density ρ_0 of C_0 is greater than a critical density $\rho_c = 1/2$. As a result, if $\rho_0 > 1/2$, then the model where the task is performed must converge to a fixed-point configuration where all states are 1, otherwise to a configuration of all 0-states.

In the literature [18], the *standard formulation* of DCT consists of determining density by means of a binary, one-dimensional CA, with its performance, or *efficacy* quantified as a percentage of correct density classifications, in a set of ICs, be it the entire set of 2^n ICs, or a sample set of generated by some known density distribution, usually binomial or uniform.

2.2 Majority Rule

The *majority rule* (MR, for short) concerns a political decision-making principle based on the will of the majority, in which the winner is chosen by the larger fraction of the individuals in a population [13].

Computational studies on MR have already been addressed in the literature (see, [3, 8] and [10]) such as in the *majority rule model* (MRM), from the field of opinion dynamics [19].

The MRM is an ABM that was first introduced in [7] to describe public debates, and can be formulated as a collection of n agents, where each one takes on a discrete opinion ± 1, and can interact with all other agents, through a complete graph, at each time step t. In this model, there are two actions to be performed. In the *first action*, n' agents are randomly selected. In the *second action*, all agents adopt the most frequent opinion within n'. Each agent might be seen as having global access to the situation in the collection, for it is connected with all others.

With this connectivity pattern, consensus is always achieved. However, this occurs only when n is odd, because there are no ties regarding the opposite opinions in the collection [19].

2.3 Circulant Graphs

Circulant graphs have been a widely covered topic in the literature [1], as they have been employed in many technical fields [12, 17]. The likely reason for such an interest is that, circulant graphs as a connectivity configuration, outperforms other topologies [12] in terms of delay, connectivity and survivability.

The family of *circulant graphs* is one of the most comprehensive families [16] of highly symmetric graphs [1], both vertex- and edge-symmetric. A graph is *vertex-symmetric* if every pair of vertices is similar, while it is *edge-symmetric* if every set of two points of an edge x is the set of points of another edge y.

Before the definition of a circulant graph, let us define an *undirected simple graph* G as an ordered pair (V, E), where V is a set of *vertices* $i \in V = \{0, ..., n - 1\}: n \in \mathbb{Z}$ and E is a set of *edges* $e_{ij} : i, j \in V$, defined by means of a mapping $f_E : V^2 \rightarrow E$, with $e = f_E(i, j)$. The number of vertices in V, denoted n, is called the *order* of G, while the *degree* d_i of vertex $i \in V$ is the number of edges said incident to i. Thus, the *average degree* $\langle d \rangle$ is the sum of all individual degrees $\sum_{i \in V} d_i$ divided by the order n.

From that, a *circulant graph* $C_{n,J}$ is generated given an order n and a set of integers $J = \{j_1, ..., j_m\} : 0 < j_1 < ... < j_m < (n + 1)/2$, from which the edges in E are defined based on the adjacencies $(i \pm j_1, i \pm j_2, ..., i \pm j_m)$ mod n, for a vertex $i \in V$. The integer j is called *jump*, while the set J is called *jump sequence* [12].

By the previous definition, there is no guarantee that G is always a connected circulant graph for any given sequence of jumps [12]. According to [1], a not-connected circulant graph occurs because the *greatest common divisor* (g.c.d.) for $(n, \{j_1, j_2, ..., j_m\})$ is higher than 1.

3   DCT Formulation

In this work, we propose a DCT formulation as a model of simple agents that seek to achieve consensus about the majority state of a binary sequence, by collectively performed local actions.

In this ABM, given a binary sequence C_0 for which the binary density ρ_0 is to be determined and a jump sequence J of a circulant graph $C_{n,J}$ is assumed as the connectivity pattern:

1. Each agent is generated for each position $i \in [0, n - 1]$ in C_0. From that, each new agent has the position i as its identity and the state $s_i \in C_0$ as its internal state. Also all the agents have global access to the order n and the jump sequence J properties, as these are essential information for determining their neighborhood N_i.
2. Each neighborhood N_i of an agent i is determined by the application of the following operations over each jump $j \in J = \left\{ j_1, \ldots, j_{\langle d \rangle /2} \right\}$:

$$\forall j \in J, (i - j)\%n \tag{1}$$

$$\forall j \in J, (i + j)\%n \tag{2}$$

3. Each internal state s_i is updated according to the *majority rule*, that is, the agent must access the state of each neighbor, thus obtaining the neighboring state set sN_i, and update s_i to the majority state that results from $sN_i \cup \{s_i\}$. This process is performed synchronously, during t timesteps.

The *efficacy* over DCT, denoted by ε, is obtained by the expression

$$\varepsilon = \kappa/c, \tag{3}$$

where κ is the total number of correctly classified densities in a set of ICs $C_0 = \{C_0^1, \ldots, C_0^c\}$ and c is the total number of ICs in C_0. In our case, since as $k = 2$, this entails that C_0 contains all 2^n binary ICs, with $n \in \mathbb{Z}$ being of odd order. These ICs are generated from a growing sequence of integers n^k, in the interval between 0 and $n^k - 1$, so that each integer is converted from its decimal base to the binary base. In this way, all combinations of length n between states 0 and 1 are generated.

4   Connectivity Influence on Performance

In order to verify the influence of connectivity on the performance of our DCT formulation, we introduce the idea of a *connectivity space*. The *connectivity space* can be defined as set $C(n)$ that contains every distinct circulant graph $C(n, J)$, whose average degree $\langle d \rangle$ belongs to a finite discrete interval $\langle d \rangle = [4, (n - 1) - 2]$.

An important point to be considered when exploring $C(n)$ is that for a same given order n and average degree $\langle d \rangle$ there are circulant graphs with distinct edges arrangements, depending on a the jump sequence J. Thus, we assume that for each average degree $\langle d \rangle$ there is a subset $C(n, \langle d \rangle) \subset C(n)$ containing all of these distinct circulant graphs. More exactly, $C(n, \langle d \rangle)$ contains each sequence J that enables the construction of the respective circulant graph $C(n, J)$.

The illustration given in Fig. 1 expresses this situation in the context of the space associated with $C(7)$; more precisely, the figure displays all distinct circulant graphs of order 7 and average degree 4, given by $C(7, 4) = \{C_{7,\{1,2\}}, C_{7,\{1,3\}}, C_{7,\{2,3\}}\}$. Naturally, as the interval increases as a function of n, the number of intermediate average degrees and their respective sets of circulant graphs also increases. Thus, by following this idea, the next spaces of connectivity to be constructed are $C(9)$, which contains two possible central sets of circulant graphs ($\langle d \rangle = 4$ and $\langle d \rangle = 6$), $C(11)$ containing three sets ($\langle d \rangle = 4$, $\langle d \rangle = 6$ and $\langle d \rangle = 8$), and so on.

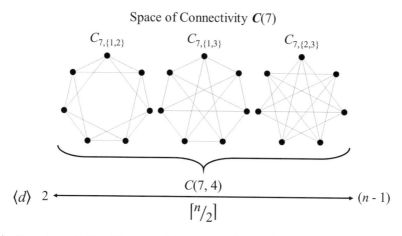

Fig. 1. Formation of all valid connectivity patterns for $C(n)|n = 7$ defined over the average degree in the interval $2 < \langle d \rangle < (n - 1)$.

Each set $C(n, \langle d \rangle) \subset C(n)$ can be generated from two steps to be performed. First, given an odd-valued order n, a sequence of integers $Q = \{1,\ldots, \lfloor n/2 \rfloor\}$ is generated. Second, tuples of length $\langle d \rangle /2$ are generated from combinations of the integers in Q. These tuples are treated in order to assume only distinct integers, in crescent order, de duplicates are eliminated. This results in all possible distinct jump sets $J = \{j_1, \ldots, j_{|J|=\langle d \rangle /2}\}$.

In our experiments, we have explored the full connectivity space $C(n)$ for each order $n \in [7, 17]$ over each average degree $\langle d \rangle$ in the interval $[4, (n - 1) - 2]$. All binary sequences in C_0 were classified and majority rule applied for $2n$ timesteps.

The results are presented in Fig. 2, where the horizontal axis of the plot represents the arrangement of all *sets of circulant graph* $C(n, \langle d \rangle)$ in each space $C(n)$, according to their *average degree* $\langle d \rangle$, and the vertical axis represents the *efficacy* values $\varepsilon(\%)$ that can be achieved by the model. Here, from the assumption of distinct connectivity patterns, the model's performance is depicted as two series of values for ε, one at the top (*triangles*) and the other at the bottom (*squares*). Thus, while the former displays the *best performance values* (ε-Max) that the model achieves for a set of circulant graphs $C(n, \langle d \rangle)$, the latter shows their counterparts, that is, the *worst values* ε-Min, in the corresponding set.

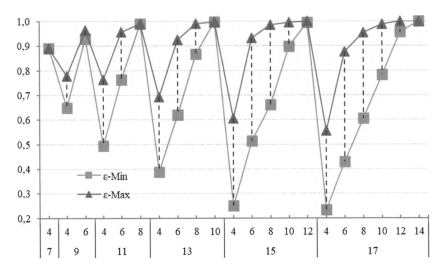

Fig. 2. Performances of the ABM model in the space $C(n)|n \in [7, 17]; \langle d \rangle = [4, (n-1) - 2]$: best values (upper plot - ε-Max) and worst values (lower plot - ε-Min).

Our first observation regarding the efficacy values presented in Fig. 2 is a phase transition at the initial average degrees when considering the interval $\langle d \rangle = [4, (n-1) - 2]$, characterized by a hop in efficacy values from one set of circulant graphs to another $C(n, \langle d \rangle)$. From that, we might understand that increasing the average degree while keeping the same order leads the model to a compatible increasing in efficacy. This can be observed by comparing the efficacies of the model regarding the sets of circulant graphs $C(13,4)$ vs. $C(13, 6)$, $C(15, 4)$ vs. $C(15, 6)$ and $C(17, 4)$ vs. $C(17, 6)$. In all these cases, the respective maximal efficacies ε-Max increase from ~69% to ~92%, ~60% to ~93% and ~55% to ~88%.

Notice that Fig. 2 displays points in which the efficacy values are the same for both ε-Min and ε-Max. This occurs for the graphs of the sets $C(7, 4)$, $C(11, 8)$, $C(13, 10)$ and $C(17, 14)$. An evident characteristics that these graphs share is that their order n is a prime number and their average degree $\langle d \rangle$ is located at the end of the interval, at $(n-1) - 2$. This is an aspect that should require further investigation so as to understand why that was so.

We have tested the isomorphism relation between the circulant graphs in sets $C(7, 4)$ and $C(9, 6)$. From that, we can conclude that the connectivity patterns that take the formulation to the same efficacy value are isomorphic, while those with different values are non-isomorphic. For example, the circulant graphs $C_{(9, J=\{1,2\})}$ and $C_{(9, J=\{1,4\})}$ are isomorphic, as they both result in $\varepsilon = 0.6484$. The same occurs with $C_{(9, J=\{1,3\})}$ and $C_{(9, J=\{2,3\})}$, resulting in $\varepsilon = 0.7773$.

Given such results, we performed a second set of test-cases to confirm whether this kind abrupt transition on efficacy might also occur given a connectivity space of higher order, defined in $C(55)$, over a half the average degree interval, with $\langle d \rangle = [4, 28]$, and a partial set C_0 of 4×10^4 binary sequences randomly generated. In each set $C(55, [4, 28]) \subset C(55)$, 10 distinct circulant graphs were arbitrarily taken, and the same set of binary sequences C_0 was used in each efficacy calculation given a circulant graph in $C(55, \langle d \rangle)$. From that, by the best graphs (ε-Max) from each set $C(55, \langle d \rangle)$: $\langle d \rangle \in [4, 28]$ we could confirm the pattern seen in the previous experiments. The resulting sequence of efficacy values corresponding to each set of circulant graphs is ε-Max-$C(55, \langle d \rangle) = \{0.0360, 0.5097, 0.8741, 0.9216, 0.9408, 0.9466, 0.9591, 0.9601, 0.9662, 0.9708, 0.9757, 0.9804, 0.9832\}$. Here, we can observe the abrupt transitions in efficacy for $C(55, 4)$ vs. $C(55, 6)$ and $C(55, 6)$ vs. $C(55, 8)$. Therefore, we can confirm that the efficacy increases proportionally to the average degree after these transitions, when $\langle d \rangle \geq 8$.

Finally, it is interesting to note that the main message all results we presented conveys are within the same level of performance of the best currently known cellular automata rules in the literature. For the record, in [20] the two best rules presented led to efficacies around 89%.

5   Concluding Remarks

Considering the points raised in the previous section, it seems clear that the variation of the average degree in relation to the order and the connectivity pattern has direct influence on the model's performance. Therefore, in order for this model to achieve high performance, it suffices that the connectivity pattern, corresponding to a circulant graph, be at the center of the average degree interval, in the central region of the connectivity space, so that around 90% of performance can be achieved in the DCT. Moreover, good performances, of at least 80%, can also be obtained by lowering the degree, from the central value (i.e., $\lceil n/2 \rceil$), downward to the first half of the interval, i.e., $\lceil n/2 \rceil / 2$.

Acknowledgements. We are grateful to financial support from CAPES (Coordenação de Aperfeiçoamento de Pessoal de Nível Superior - Brazil) for a grant to the first author, and for the following projects also funded by CAPES: STIC-AmSud project CoDANet 88881.197456/2018-01, and PrInt project no. 88887.310281/2018-00.

References

1. Boesch, F., Tindell, R.: Circulants and their connectivities. J. Graph **8**(4), 487–499 (1984)
2. Bonabeau, E.: Agent-based modeling: methods and techniques for simulating human systems. Proc. Natl. Acad. Sci. U.S.A. **99**, 7280–7287 (2002)
3. Chen, P., Redner, S.: Majority rule dynamics in finite dimensions. Phys. Rev. **71**, 036101 (2005)

4. Das, R., Crutchfield, J.P., Mitchell, M., Hanson, J.E.: Evolving globally synchronized cellular automata. In: Proceedings of the Sixth International Conference on Genetic Algorithms, pp. 336–343 (1995)
5. de Oliveira, P.P.B.: On density determination with cellular automata: results, constructions and directions. J. Cell. Automata 9(5–6), 357–385 (2014)
6. Durfee, E.H.: Distributed problem solving and planning. In: ECCAI Advanced Course on Artificial Intelligence. ACAI 2001: Multi-Agent Systems and Applications, pp. 118–149 (2001)
7. Galam, S.: Application of statistical physics to politics. Phys. A 274, 132–139 (1999)
8. Gärtner, B., Zehmakan, A.N.: Majority model on random regular graphs. In: Latin American Symposium on Theoretical Informatics, vol. 10807, pp. 572–583 (2018)
9. Kari, J., Le Gloannec, B.: Modifed traffic cellular automaton for the density classification task. Fundam. Inf. 116(1–4), 141–156 (2012)
10. Krapivsky, P.L., Redner, S.: Dynamics of majority rule in two-state interacting spin systems. Phys. Rev. Lett. 90, 238701-1–238701-4 (2003)
11. Land, M., Belew, R.K.: No perfect two-state cellular automata for density classification exists. Phys. Rev. Lett. 74(25), 5148 (1995)
12. Lu, R.: Fast methods for designing circulant network topology with high connectivity and survivability. J. Cloud Comput. Adv. Syst. Appl. 5, 5 (2016)
13. McGann, A.J.: The tyranny of the super-majority: how majority rule protects minorities. UC Irvine – Center for the Study of Democracy (2002)
14. Mitchell, M., Crutchfield, J.P., Hraber, P.T.: Evolving cellular automata to perform computations: mechanisms and impediments. Physica D 75, 361–391 (1994)
15. Mitchell, M.: Computation in cellular automata: a selected review. In: Gramb, T., Bornholt, S., Grob, M., Mitchell, M., Pellizzari, T. (eds.) Non-standard Computation, pp. 95–140. Wiley, Weinheim (1998)
16. Munir, M., Nazeer, W., Shahzadi, Z., Kang, S.M.: Some Invariants of Circulant Graphs. Symmetry 8, 134 (2016)
17. Newman, M.: The structure and function of complex networks. SIAM Rev. 45, 167–256 (2003)
18. Packard, N.H.: Adaptation towards the edge of the chaos. In: Kelso, J.A.S., Mandell, A.J., Shlesinger, M.F. (eds.) Dynamic Patterns in Complex Systems, pp. 293–301 (1988)
19. Sîrbu, A., Loreto, V., Servedio, V.D.P., Tria, F.: Opinion dynamics: models, extensions and external effects. In: Participatory Sensing, Opinions and Collective Awareness, pp. 363–401 (2016)
20. Wolz, D., de Oliveira, P.P.B.: Very effective evolutionary techniques for searching cellular automata rule spaces. J. Cell. Automata 3(4), 289–312 (2008)

Fault Detection Mechanism of a Predictive Maintenance System Based on Autoregressive Integrated Moving Average Models

Marta Fernandes[1(✉)] , Alda Canito[1] , Juan Manuel Corchado[2] ,
and Goreti Marreiros[1]

[1] GECAD - Research Group on Intelligent Engineering and Computing
for Advanced Innovation and Development, Polytechnic of Porto (ISEP/IPP),
Porto, Portugal
{mmdaf, alrfc, mgt}@isep.ipp.pt
[2] BISITE Research Centre, University of Salamanca (USAL), Salamanca, Spain
corchado@usal.es

Abstract. The industrial world is amid a revolution, titled Industry 4.0, which entails the use of IoT technologies to enable the exchange of information between sensors, industrial machines and end users. A major issue in many industrial sectors is production inefficiency, with process downtime representing a loss for companies. Predictive maintenance, whereby maintenance is performed only when needed and before a failure occurs, has the potential to substantially reduce costs. This paper describes the fault detection mechanism of a predictive maintenance system developed for the metallurgic industry. Considering no previous information about faults is available, learning happens in an unsupervised manner. Imminent faults are predicted by estimating autoregressive integrated moving average models using real-world sensor data obtained from monitoring different machine components and parameters. The models' outputs are fused to assess the significance of an anomaly (or anomalies) along the time domain and determine how likely a fault is to occur, with alarms being issued when the prospect of a fault is high enough.

Keywords: Predictive maintenance · Anomaly detection · ARIMA models · Sensor data

1 Introduction

The industrial world is in the midst of a new revolution fueled by the use of Cyber Physical Systems that interact and communicate with each other and with human beings, leading to an increase in productivity, efficiency and security, while reducing waste and improving the work experience [1, 2]. This transformation, titled Industry 4.0, entails the integration of IoT technologies into the production cycle, enabling the exchange of information between sensors, industrial equipment and end users. This creates an environment where it is possible for a worker, in any part of a factory, to access a real-time digital representation of the production process and current statuses of the machines in use [1, 2]. Although companies have been adopting modern

F. Herrera et al. (Eds.): DCAI 2019, AISC 1003, pp. 171–180, 2020.
https://doi.org/10.1007/978-3-030-23887-2_20

technologies as they become available, in many industrial sectors small and medium-sized companies are still lagging behind [1, 3].

The information acquired with the digitalization of the industrial sector can be used by stakeholders to create new business opportunities and improve a company's competitiveness. A major issue in many industrial sectors is production inefficiency, with process downtime representing a loss for the company [4]. Optimizing process operations is crucial to maximize a company's productivity and limit consumption, which results in a reduction of costs and environmental impact [1, 4].

Machine downtime and production costs are strongly dependent on a company's maintenance policy [3, 5]. To maximize machine uptime, problems must be detected and corrected before machines reach the point of failure. Machine failures can not only result in greater downtime, when compared to the downtime required by preventive interventions, but they can also cause greater damage in the machines, as well as damage articles in production at the time the fault occurs [6]. Proactive maintenance practices combine preventive and predictive maintenance to anticipate problems, repairing or replacing components after a certain level of deterioration has been detected, instead of performing repairs after a fault occurs [7]. In this way, a predictive maintenance system aims to maximize the interval between interventions on the machines, while minimizing unscheduled repairs.

The implementation of predictive maintenance relies on intelligent decision methods and requires operational and process data. This information can be acquired by employing IoT technologies and principles of Cyber Physical Systems to interface with legacy systems and set-up unobtrusive sensors networks that monitor the condition of industrial equipment [8, 9].

Processing sensor data is a difficult task due to the high volume of data generated, the speed at which it is generated, arriving in streams that can grow indefinitely, and the dynamics of a continuously changing environment [10, 11]. Additionally, this data consists in time series: a sequence of data points recorded at specific times, where the temporal continuity is the predominant factor to consider [12]. Machine Learning and statistical models can be used to draw insights from this data [12–14], predicting outcomes that support decision-making and help organizations improve their operations and competitiveness [6, 15].

In this paper, we present the fault detection mechanism of a predictive maintenance system developed in the scope of The Industrial Enterprise Asset Value Enablers (InValue) project [16]. The InValue project has the purpose of promoting the transition to proactive maintenance practices in a metallurgical company specialized in the production of custom precision parts. The fault detection mechanism uses multimodal sensor data to train autoregressive integrated moving average (ARIMA) models [17]. The models' outputs are then fused and used to determine how likely a fault is to occur.

This document is organized as follows: (1) Introduction, (2) Context, (3) Predictive Models, (4) Fault Detection Mechanism and (5) Conclusions.

2   Context

The data used to train the ARIMA models consists in real-world data, obtained from monitoring four CNC machines in a mechanical metallurgy factory. This company specializes in precision parts production, using raw materials, such as aluminium, steel, bronze and technical plastics to produce custom parts for industry clients, particularly the automotive industry. The data was acquired from two lathes and two vertical mills by installing sensors and interfacing with the machines' firmware to consume the information provided by the inner sensors [18].

The data was initially characterized by forty-seven attributes, four of which were captured from external sensors. After undergoing an exploratory phase and a process of feature selection, the number of features of interest was reduced to thirty-two. Of these, seven were considered useful for prediction purposes, namely: spindle load, spindle rotations per minute (RPM), coolant level, vibration for axis X, Y and Z, and noise. This process is described in detail in [19].

The 'spindle load' refers to the energy outputted by the vector drive that powers the spindle's motion, while the 'spindle RPM' measures its rotational speed. Both features have well defined maximum theoretical values that are different for each machine model. The spindle can safely sustain loads up to 200% of the maximum value for at most thirty minutes (the greater the load, the shorter the duration). Both the spindle load and the spindle rpm are strong indicators of problems in the machines.

The vibration and the noise are also good predictors that something might be wrong with a machine. A faulty component or an abnormal event will cause changes to the typical vibration and noise patterns of a machine. This might occur abruptly, in case of a sudden malfunction in one or more of the machine's components, or it might happen slowly as the components deteriorate.

Another important variable is the coolant fluid, a water-oil emulsion used to cool the cutting tool and the workpiece. A low level of coolant fluid has implications in the wear and tear of the cutting tool and might be responsible for changes in the metal-lurgical characteristics of the workpiece material.

These and other variables of interest were used to define a rule-based model that determines the actions the system must take when certain conditions are met [19]. This rule-based system has the purpose of complementing the fault detection mechanism presented in this paper.

Since the data comes from different sources and has different modalities, it is captured at different rates, as shown in Table 1. The constant monitorization of the machines, coupled with the high rate of acquisition of some features, generates vast amounts of data. Moreover, due to bandwidth and software limitations, the information is acquired at irregular intervals. Considering most time series analysis methods expect regularly spaced data points, the data was downsampled to 5-min intervals, which also reduced its volume substantially.

Table 1. Current data acquisition rates.

Data	Acquisition rate
Spindle load (machine sensor)	0.2 Hz
Spindle RPM (machine sensor)	0.2 Hz
Coolant level (machine sensor)	0.2 Hz
Noise (external sensor)	100 Hz
Vibration (external sensors)	100 Hz

An important aspect of this case study is that no information about faults is available. The CNC machines of the metallurgical company involved in the project fail very rarely due to a strict schedule of preventive maintenance and to the fact that the company only produces small series of custom parts, which wears down the machines less than mass production would. This constrains the type of methods that can be used to model the data because the prediction of faults can't be approached as a supervised learning problem.

3   Predictive Models

Since data concerning problems in the machines isn't available, the prediction of faults was framed as an anomaly detection problem, using classical time series methods to build forecasting models. In time series there is temporal dependence between the data points, meaning the order of the data is critical. Furthermore, the analysis performed on time series should consider the data may have internal patterns, such as autocorrelation and seasonality, among others [20].

The fault detection mechanism here described uses autoregressive integrated moving average models to model the autocorrelations in the data. ARIMA models combine autoregressive models, moving average models and differencing operations. However, not all of these components need to be present. If a model only possesses autoregressive terms, it's called an AR model. Similarly, if only moving average terms are present, it is an MA model. Differencing operations might also not be necessary, in which case the resulting model will be referred to as ARMA.

As the name implies, an autoregressive model predicts a variable's present value by using a linear combination of its past values (or lagged values). Equation 1 refers to an autoregressive model of order p ($AR(p)$), where p indicates how many lagged values of the variable of interest are used to predict the present value [21].

$$Y_t = \delta + \sum_{i=1}^{p} \phi Y_{t-i} + \varepsilon_t \tag{1}$$

Likewise, moving average models use linear combinations of past forecast errors to predict a variable's present value. An $MA(q)$ model is a moving average model of order q, as shown in Eq. 2 [21].

$$Y_t = \mu + \varepsilon_t + \theta_1 \varepsilon_{t-1} + \theta_2 \varepsilon_{t-2} + \ldots + \theta_q \varepsilon_{t-q} \tag{2}$$

ARIMA models, like most statistical forecasting methods, assume a time series is (weakly) stationary or can become stationary by applying some mathematical transformation to the data. A time series is stationary when its statistical properties, such as mean, variance and autocorrelation, remain constant across time. Differencing operations are one way of making a time series stationary. The I in ARIMA refers to the order of integration of a time series, that is, the number of differencing operations required to make the series stationary. For example, I(1), which represents the first difference, is defined by Eq. 3 [21]:

$$Y'_t = y_t - y_{t-1} = (1 - B)Y_t, \text{ where } By_t = y_{t-1} \tag{3}$$

Since ARIMA models are a combination of AR and MA models, they rely on linear combinations of a variable's past values and past forecast errors to model present values and forecast future values. Equation 4 denotes a non-seasonal ARIMA(p, d, q) model, where p is the order of the autoregressive component, d is the number of differencing operations necessary to make the series stationary and q is the order of the moving average component [21].

$$\left(1 - \sum_{i=1}^{p} \phi_i B^i\right)(1 - B)^d y_t = \delta + \left(1 + \sum_{i=1}^{q} \theta_i B^i\right) \varepsilon_t \tag{4}$$

ARIMA models can also be seasonal, combining non-seasonal AR and MA terms with seasonal AR and MA terms in a multiplicative model. Non-seasonal and/or seasonal differencing can also be performed, if necessary. A seasonal ARIMA model is represented by the notation: ARIMA(p, d, q) $\times$ (P, D, Q)m, where 'm' is the number of time periods until the pattern in the data repeats again [21].

ARIMA models are applicable only to univariate time series data. For this reason, it was necessary to build one model for each of the variables and machines presented in Sect. 2.

Model selection was performed by estimating several models with different parameters for each variable of interest. The validity of each model was assessed through analysis of the residuals. An ARIMA model is valid if it has correctly modeled the patterns in the data, which can be verified by analyzing certain properties of the model's residuals. It is essential that the residuals exhibit no correlation and their mean should be zero. Ideally, the residuals should also have constant variance and be normally distributed, but while these properties make the calculation of prediction intervals easier, they are not essential [21]. Figure 1 shows the residuals of a model with mean approximately equal to zero, no significant correlation between the residuals and constant variance. However, as can be seen, the residuals do not follow a Gaussian distribution.

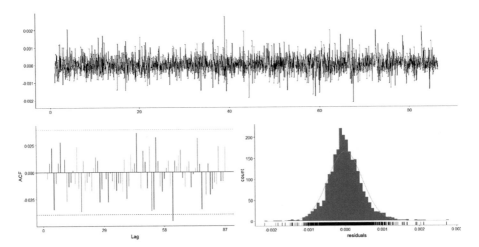

Fig. 1. Residuals of an ARIMA model.

After validating the models, the "best" model was chosen by minimizing the Akaike Information Criterion (AIC) [21, 22]. The AIC can be used to compare the quality of different forecasting models. Since it rewards a model's goodness of fit, while penalizing its complexity, the AIC performs a trade-off between overfitting and underfitting [22].

Figure 2 shows training and test vibration data overlaid with 95% and 80% prediction intervals derived from a seasonal ARIMA model. The root mean squared error (RMSE) on the test data is equal to 0.00063.

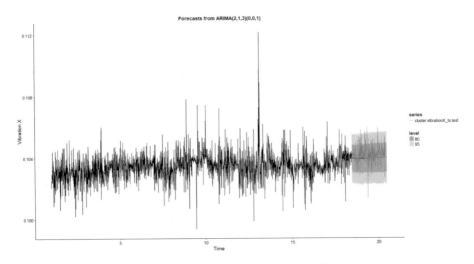

Fig. 2. 95% and 80% prediction intervals from an ARIMA(2,1,3)(0,0,1) model.

Due to the scale of the plot, the model's point forecasts appear to differ greatly from the test data but in fact their mean value only differs 0.24% from the mean value of the test data. Nonetheless, point forecasts are oftentimes incorrect because the distribution of a real-world time series is not stationary, that is, its statistical properties change over time. As such, prediction intervals represent the uncertainty of the forecasts and point forecasts are of little interest without them. As can be observed in Fig. 2, most of the test data falls within the boundaries of the 95% prediction interval and the 80% prediction interval captures the central tendency of the data quite well. The prediction intervals are far more useful to detect anomalies in the machines than the point forecasts. Hence, the proposed fault detection mechanism applies the 95% prediction interval.

4 Fault Detection Mechanism

The predictive models built for the features of interest were used in the development of a fault detection mechanism whose purpose is issuing alarms whenever a fault is likely to occur. For each feature, this is achieved by comparing new data with the respective 95% prediction interval. If a previously unseen value falls outside the prediction interval it is considered an anomaly. However, an anomaly doesn't have much significance by itself, it's the accumulation of anomalies over a given time period that indicates the possibility of a malfunction. That occurrence is even more significant if several anomalies occur concurrently in more than one variable.

For a single variable, a 30-min moving average of anomaly occurrences is calculated, and an alarm is issued if the average exceeds a certain value. The default value is 0.85 – that is, at least 85% of the values in the 30-min window must be outside the 95% prediction interval – but this threshold can be changed according to the end user's needs. Table 2 shows an example of the average number of anomalies for three different variables. In this case, an alarm would be issued for the spindle load but not for the other variables.

Table 2. Example of average of anomaly occurrences in a 30-min windows.

Time	Spindle load	Spindle RPM	Vibration
0 min	Anomaly	Normal	Anomaly
5 min	Anomaly	Normal	Normal
10 min	Normal	Normal	Normal
15 min	Anomaly	Normal	Anomaly
20 min	Anomaly	Normal	Anomaly
25 min	Anomaly	Normal	Normal
30 min	Anomaly	Normal	Normal
Avg. anomalies:	0.857	0	0.428

It is not enough to consider variables independently, since a fault might involve more than one variable. To account for this possibility, the outputs of each ARIMA model (for a given machine) are combined and a correlation matrix of the occurrence of anomalies is calculated along a 30-min moving window, using the Pearson correlation coefficient. If the correlation coefficient between two variables is greater than 0.7, the anomalies in each variable are individually analyzed using the method described for fault detection in single variables. However, the more variables display anomalies concurrently, the more serious a malfunction might be; therefore, alarms or notifications should be sent out more urgently. This situation was addressed by defining the alarm threshold in proportion to the number of variables that display correlation of anomalies. The more variables are involved, the lower the alarm threshold will be, as defined by Eq. 5.

$$threshold = 1 - n\_var\_analysed \frac{0.7}{n\_var\_tot} \tag{5}$$

In the equation, "$n\_var\_analysed$" refers to the number of variables with correlated anomalies and "n_var_tot" is the number of total variables, which, in this case, is seven (per machine). It was defined that the alarm threshold must have a minimum value of 0.3. If none of the variables in consideration has an average of anomalies greater than 0.3, their occurrence isn't considered significant. As such, no alarm will be issued if each of the variables with correlated anomalies had less than 30% of anomalies in the previous 30 min, but if the average of anomaly occurrences of at least one of them exceeded the threshold defined in Eq. 5, an alarm will be sent out.

5   Conclusion

In this paper, we proposed a fault detection mechanism for a predictive maintenance system, which relies on the forecasts of autoregressive integrated moving average models to detect anomalies in CNC machines. Predictive maintenance systems aim to predict faults in industrial equipment to avoid unnecessary downtime and expensive repairs. Due to the lack of data regarding previous faults in the machines, it wasn't possible to use supervised learning techniques to accurately predict the occurrence of malfunctions. As such, ARIMA models, a classical time series method, were used to produce prediction intervals from the sensor data obtained from monitoring the condition of the machines.

The detection of anomalies is performed by comparing new data to the 95% prediction intervals. Any data that falls outside the bounds of an interval is considered an anomaly. Since isolated anomalies may not be indicative of problems in a machine, the likelihood of a fault is assessed by a 30-min moving average of anomaly occurrences. Alarms are issued if the average number of anomalies of a single variable exceeds a user-defined value (default = 0.85), but the incidence of simultaneous anomalies across different variables is also considered, using correlation-based decision fusion.

The mechanism described in this paper represents an initial approach to the problem of performing predictive maintenance when no labeled data is available.

Future work will include trying out other learning methods and mechanisms to perform a comparative analysis of the results obtained.

Acknowledgements. The authors wish to acknowledge the Portuguese funding institution FCT - Fundação para a Ciência e a Tecnologia for supporting their research through project UID/EEA/00760/2019 and Ph.D. Scholarship SFRH/BD/136253/2018.

References

1. Evans, P.C., Annunziata, M.: Industrial internet: pushing the boundaries of minds and machines (2012)
2. Boyes, H., Hallaq, B., Cunningham, J., Watson, T.: The industrial internet of things (IIoT): an analysis framework. Comput. Ind. **101**, 1–12 (2018)
3. Holmberg, K., Adgar, A., Arnaiz, A., Jantunen, E., Mascolo, J., Mekid, S. (eds.): E-maintenance. Springer, London (2010)
4. Williamson, J.: Unplanned downtime affecting 82% of businesses. https://www.themanufacturer.com/articles/unplanned-downtime-affecting-82-businesses/. Accessed 25 Jan 2019
5. Aboelmaged, M.: Predicting e-readiness at firm-level: an analysis of technological, organizational and environmental (TOE) effects on e-maintenance readiness in manufacturing firms. Int. J. Inf. Manag. **34**, 639–651 (2014)
6. Selcuk, S.: Predictive maintenance, its implementation and latest trends. In: Proceedings of the Institution of Mechanical Engineers, Part B: Journal of Engineering Manufacture, vol. 231, pp. 1670–1679 (2017)
7. Muller, A., Marquez, A., Iung, B.: On the concept of e-maintenance: Review and current research. Reliab. Eng. Syst. Saf. **93**, 1165–1187 (2008)
8. Al-Fuqaha, A., Guizani, M., Mohammadi, M., Aledhari, M., Ayyash, M.: Internet of things: a survey on enabling technologies, protocols, and applications. IEEE Commun. Surv. Tutor. **17**, 2347–2376 (2015)
9. Lee, J., Jin, C., Bagheri, B.: Cyber physical systems for predictive production systems. Prod. Eng. **11**, 155–165 (2017)
10. Krempl, G., Žliobaite, I., Brzeziński, D., Hüllermeier, E., Last, M., Lemaire, V., Noack, T., Shaker, A., Sievi, S., Spiliopoulou, M., Stefanowski, J.: Open challenges for data stream mining research. ACM SIGKDD Explor. Newsl. **16**, 1–10 (2014)
11. Gama, J.: A survey on learning from data streams: current and future trends. Prog. Artif. Intell. **1**, 45–55 (2012)
12. Gupta, M., Gao, J., Aggarwal, C., Han, J.: Outlier detection for temporal data: a survey. IEEE Trans. Knowl. Data Eng. **26**, 2250–2267 (2014)
13. Esling, P., Agon, C.: Time-series data mining. ACM Comput. Surv. **45**, 12 (2012)
14. Ahmed, N.K., Atiya, A.F., Gayar, N.E., El-Shishiny, H.: An empirical comparison of machine learning models for time series forecasting. Econom. Rev. **29**, 594–621 (2010)
15. Jardine, A.K.S., Lin, D., Banjevic, D.: A review on machinery diagnostics and prognostics implementing condition-based maintenance. Mech. Syst. Signal Process. **20**, 1483–1510 (2006)
16. InValuePT. http://www.invalue.com.pt/
17. Box, G.E.P., Jenkins, G.M.: Time series analysis: forecasting and control. Holden-Day (1970)

18. Canito, A., Fernandes, M., Conceição, L., Praça, I., Santos, M., Rato, R., Cardeal, G., Leiras, F., Marreiros, G.: An architecture for proactive maintenance in the machinery industry. In: Advances in Intelligent Systems and Computing, pp. 254–262. Springer (2017)
19. Fernandes, M., Canito, A., Bolón-Canedo, V., Conceição, L., Praça, I., Marreiros, G.: Data analysis and feature selection for predictive maintenance: a case-study in the metallurgic industry. Int. J. Inf. Manage. **46**, 252–262 (2018)
20. Shumway, R.H., Stoffer, D.S.: Time Series Analysis and Its Applications. Springer, Cham (2017)
21. Hyndman, R.J., Athanasopoulos, G.: Forecasting: Principles and Practice, 2nd edn. OTexts, Melbourne (2018)
22. Burnham, K.P., Anderson, D.R.: Multimodel inference: understanding AIC and BIC in model selection. Sociol. Methods Res. **33**, 261–304 (2004)

Object Recognition: Distributed Architecture Based on Heterogeneous Devices to Integrate Sensor Information

Jose-Luis Poza-Lujan(✉), Juan-Luis Posadas-Yagüe, Eduardo Munera, Jose E. Simó, and Francisco Blanes

University Institute of Control Systems and Industrial Computing (ai2), Universitat Politècnica de València (UPV), Camino de Vera, s/n., 46022 Valencia, Spain
{jopolu,jposadas,emunera,jsimo,fblanes}@ai2.upv.es

Abstract. Object recognition is a necessary task for many areas of technology, such as robot navigation or the intelligent reconstruction of environments in order a robot can interact with these objects. This article presents an architecture that integrates distributed heterogeneous information to recognise objects. The architecture uses devices that can process sensory data locally to send information to other devices. In order to perform data processing actions, the devices must have a layer of intelligent connectivity. These devices are called Smart Resources that offer to the distributed nodes the processed sensor data by means of information services. To validate the architecture, a system with two smart resources, equipped with different sensors, has been implemented. Experiments carried out show that it is better to select objects as soon as possible to improve the object recognition rate. Consequently, in the distributed system, devices should, when possible, deliver the process in advance.

Keywords: Smart environment · Distributed systems · Distributed architectures · Object detection · Information integration

1 Introduction: The Environment Knowledge Acquisition Process

Smart systems usually must perform their tasks in dynamic environments with multiple features and changing conditions. Therefore, in order to provide autonomy to interact with the environment, the continuous and accurate knowledge of the environment is necessary.

Systems such as robot navigation need to know the environment to perform their tasks such as the planning of trajectories or the execution of missions [15]. Sensors can help to know the environment by detecting objects and some of their characteristics. However, when the objects detected have to be classified

© Springer Nature Switzerland AG 2020
F. Herrera et al. (Eds.): DCAI 2019, AISC 1003, pp. 181–188, 2020.
https://doi.org/10.1007/978-3-030-23887-2_21

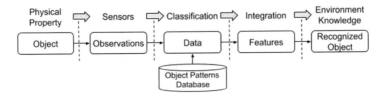

Fig. 1. Simplification of the recognition process in integration of sensory information

and recognised, a set of patterns with which to compare [10] is necessary. For example, the shape of a box can be detected by means of a 3D sensor, but the same box can have different textures, so it is also necessary other type of sensor, for example an RGB sensor, to recognise what type the box is. Therefore, using heterogeneous sensors to detect and recognise the objects placed in an environment can increase the probability of success recognising the right object. When working with heterogeneous sensors, their information must be merged, usually remotely creating sensor networks [18].

As a summary, acquiring characteristics of the environment to associate them with specific objects implies a sequence of actions shown in the Fig. 1.

The inclusion of object detection in the environment map adds a difficulty and involves the use of advanced sensors. Consequently, when there are many sensors the data fusion depends on the fusion mechanism [4]. Once a certain precision in the detection of the object and its characteristics has been achieved, it should be possible to classify the object [1]. The classification of the object requires the use of patterns in order to compare the percentage of similarity [8]. Therefore, in an object recognition system, a database of patterns is necessary.

The addition of micro-controllers and micro-processors to the sensors devices, has increased the information capacity that sensors can provide. These devices are usually called respectively smart, or intelligent sensors [17]. When the sensor includes some advanced processing and, in some occasions, actuators, some authors call them smart devices [12]. Adding communication interfaces allows smart devices to share information and, consequently increase the knowledge of the environment. The use of smart devices is growing from the environment like Smart cities [7] to the concept of Smart Objects when these devices are included into the daily life of people [9].

Consequently, the current situation is that sensors can send processed information rather than raw data. The result is that sensor networks become into distributed systems that integrate sensor information in order to take advantage of the processed information [2]. When there are different distributed devices, there are some interesting problems to solve. One of the problems is to achieve a balance between the number of devices used and the correct use of their sensors. That is, when a new device is introduced, its sensors should help increase the probability of success when detecting and recognising an object. Consequently, the composition and connection between the devices will determine how to recognise the objects. For example, two devices with an RGB sensor can recognise the same texture with a similar probability. However, the probability of success could increase by using another type of sensor which reinforces the process, for example, a thermal camera that can distinguish between different ink types.

To study how different types of devices can cooperate, the Smart Resource model used in previous researches has been used [13]. Smart resources allow high connection flexibility since the features are offered as services. Services offered depend on the available sensors and the computing capacity of each smart resource. Clients determine the necessary services, establishing a connection topology depending on their needs. For example, in the case of a smart resource that detects and identifies an object with a high probability, more information to corroborate the identified object could not be required. However, if the probability is low, the smart resource will need other measurements from other sensors that allow to increase the probability to identify successfully the object.

The aim of this paper is to present the study and implementation of a solution to integrate sensory information in order to increase certainty in the object recognition process. The Sect. 2 presents the principles and components used in the experiments. The Sect. 3 "Experiments and Results" shows the case study tested and the results obtained to recognise two similar objects. The results obtained verify how the integration of information from the services provided by the smart resources improves the accuracy to detect objects. Finally, in Sect. 4 some conclusions will be drawn.

2   System Architecture

As described in the previous section, the classification and integration steps imply the use of patterns and the decision is based on the probabilities of each pattern provided from the different sensors. The process is described in Fig. 2.

Taking into account that each object j has a specific pattern from each type of sensor i. A pattern is built with the object characteristics detected from a type of sensor. If the whole process was centralised, each device should have access to as many sensors as possible, and the patterns to compare with those sensors. The storage load of all the patterns and the processing load of all the sensors in the same device could be too much. In addition, in the case that a sensor obtains a very high probability with a specific object pattern, it would be not necessary to continue processing more sensors, unless a 100% certainty was required. Therefore, a distributed system can be an adequate and efficient solution. In this system, only when a device has a low certainty in the recognition of an object, it should request more results from other devices that have recognised that object. So, a device A needs to have only the patterns of the sensors used. The device A should be able to consult a device B about an object, in order to reinforce the probability of the recognised object. In order to distribute the object recognition process, a system based on distributed intelligent devices, called Smart Resources, has been developed.

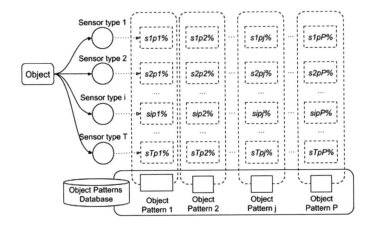

Fig. 2. A framework of object recognition problem solved based on patterns

The smart resource model is described in [11]. A smart resource is an extension of a smart device, that offers services to the others system elements. For example, in the system described in the next section, smart resources offer as a services, among others, their position in the map, and the probability detected for each pattern.

In order to communicate the devices, a communication system is needed that allows the subscription to specific services, offering a balanced network load. For example, in the Fig. 2, the object of pattern 1 may have associated the sensors type 1, 2, and 3. But if there is a device that only has a sensor type 1 and 2, and another device with the type 3 sensor, it is convenient for both devices to send and receive information of a type of pattern and not of a specific device. The Publish-Subscribe [5] paradigm is one of the most suitable since it allows to uncouple physically the devices involved in the communication and connect each device to the information that they are interested in. Communications system used is the CKMultiPeer, described in [14], CKMultiPeer is based in the DDS model, widely used in critical distributed systems [16].

The communication system allows a device to connect to a source of information (the pattern of a specific object) from which to obtain data that can reinforce the identification of a specific object, fulfilling the requirements mentioned above.

3   Experiments and Results

3.1   System Implemented

To validate the architecture and test the operation of smart resources a system with two smart resources has been implemented. System is shown in Fig. 3. Two robots Turtlebot [6] carry on the smart resources. Any smart resource is composed by one BeagleBone [3], the corresponding sensors and one

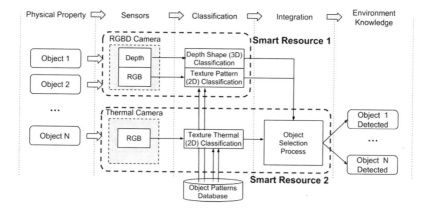

Fig. 3. Details of the system implemented to perform the experiments and the corresponding step associated with the component (top of the figure).

IEEE802.11 interface to communicate between them. Turtlebot 1 carry on the Smart Resource 1 and Turtlebot carry on Smart Resource 2. The first smart resource used, has two sensors, a deep camera to detect the geometry and an RGB camera to detect the texture. The second smart resource has only a sensor, a thermal camera that produces an RGB associated to the colour reflected. The colour of the image depends on the temperature and is directly associated with the ink composition.

The reason for using a different RGB sensor is to be able to use the same recognition algorithms (2D image), but with different patterns of the same object.

3.2   Scenario and Experiment Performed

In the proposed system shown in the Fig. 3, two specific objects are proposed to be detected and recognised by means of two different smart resources. Both objects have the same geometry (boxes) but different textures (the box of a Xtion and the box of a BeagleBone). The two boxes used to the experiments are presented in the bottom of the Fig. 4.

The experiment starts when the two robots find the box. First was tested with the Beaglebone box, and after the Xtion box. When Smart Resource 1 detects a box with a reasonable prospect of certainty (upper than 0.500) publish the estimated box position, time and certainly value in the topic 'BBBox' or 'XtionBox'. A topic is a common space to share data in a Publish-subscribe system. Smart Resource 2, receives the data of the certainly of both boxes and integrates the information with the data obtained from their sensor.

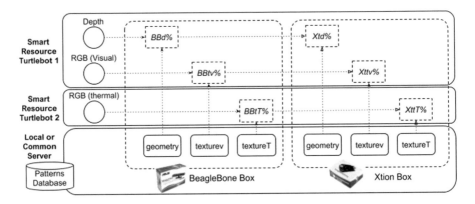

Fig. 4. Objects (boxes) used in the experiments, sensor into the smart resources, and patters used to recognise the boxes.

Table 1. Results of applying the integration method with two similar objects patterns (BeagleBone box and Asus Xtion box)

Object pattern	Geometry	Texture	Fusion	Texture	Integration
Object: BeagleBone box	Turtlebot 1			Turtlebot 2	
BeagleBone box	0.726	0.671	0.792	0.789	0.824
Asus Xtion box	0.647	0.127	0.651	0.192	0.658
Object: Xtion box	Turtlebot 1			Turtlebot 2	
BeagleBone box	0.243	0.231	0.253	0.210	0.259
Asus Xtion box	0.851	0.712	0.886	0.812	0.902

3.3   Results

In the proposed scenario, the two robots (Turtlebot 1 and Turtlebot 2) navigate until they detect the same set of objects. Both robots have a different perspective, and they are located correctly on the map. To show the process better, the results of each detected object have been shown separately. The Table 1 shows the results obtained when the data from the sensors of the first robot (Turtlebot 1) is compared with the geometries of boxes. It can be seen that both are very similar, with a certain difference favourable to the box of the BeagleBone. In the case of texture, the RGB sensor clearly detects a tendency to the correct object.

As can be seen in the tables, when the Smart Resource 2 requests the system (through the CKMultiPeer topics) the certainty of the object, the correct object is always reinforced. Data in the opposite direction, when the integration is done by Smart Resource 1, and the information of certainty is provided by Smart Resource 2, are similar. Consequently, it is possible that two uncoupled and heterogeneous systems collaborate to improve their perceptions.

4   Conclusions and Future Work

The union of sensors, micro-controllers and communications allows the implementation of intelligent distributed systems. In this paper has been presented how considering services as a communication method in a smart device, it allows to integrate information from different sensors. The experiments carried out to verify the integration of the information, increase notably the success in the detection of an object.

Based on the experiments carried out, it is convenient to test how the smart resources employees detect other objects. The paper shows the experiments performed with a system with two smart resources detecting two different objects, adding more objects and smart resources it is possible to study the cost in workload to recognise an environment. Distributing the objects characteristics to be recognised, it is possible to balance the workload to use an optimal amount of system resources, such as a processing time or communications load.

Acknowledgments. Work supported by the Spanish Science and Innovation Ministry MICINN: CICYT project PRECON-I4: "Predictable and dependable computer systems for Industry 4.0" TIN2017-86520-C3-1-R.

References

1. Azim, A., Aycard,O.: Detection, classification and tracking of moving objects in a 3D environment. In: 2012 IEEE Intelligent Vehicles Symposium (IV), pp. 802–807. IEEE (2012)
2. Chen, Y.: Industrial information integration–a literature review 2006–2015. J. Ind. Inf. Integr. **2**, 30–64 (2016)
3. Coley, G.: Beaglebone Black System Reference Manual. Texas Instruments, Dallas (2013)
4. Deng, X., Jiang, Y., Yang, L.T., Lin, M., Yi, L., Wang, M.: Data fusion based coverage optimization in heterogeneous sensor networks: a survey. Inf. Fusion **52**, 90–105 (2019)
5. Eugster, P.T., Felber, P.A., Guerraoui, R., Kermarrec, A.-M.: The many faces of publish/subscribe. ACM Comput. Surv. (CSUR) **35**(2), 114–131 (2003)
6. Garage, W.: Turtlebot, pp. 11–25 (2011). http://turtlebot.com/
7. Hancke, G.P., Hancke Jr., G.P., et al.: The role of advanced sensing in smart cities. Sensors **13**(1), 393–425 (2012)
8. Jain, A.K., Duin, R.P., Mao, J.: Statistical pattern recognition: a review. IEEE Trans. Pattern Anal. Mach. Intell. **22**(1), 4–37 (2000)
9. Lazar, A., Koehler, C., Tanenbaum, J., Nguyen, D.H.: Why we use and abandon smart devices. In: Proceedings of the 2015 ACM International Joint Conference on Pervasive and Ubiquitous Computing, pp. 635–646. ACM (2015)
10. Lim, G.H., Suh, I.H., Suh, H.: Ontology-based unified robot knowledge for service robots in indoor environments. IEEE Trans. Syst. Man Cybern.-Part A: Syst. Hum. **41**(3), 492–509 (2011)
11. Munera, E., Poza-Lujan, J.-L., Posadas-Yagüe, J.-L., Simó-Ten, J.-E., Noguera, J.F.B.: Dynamic reconfiguration of a RGBD sensor based on QoS and QoC requirements in distributed systems. Sensors **15**(8), 18080–18101 (2015)

12. Poslad, S.: Ubiquitous Computing: Smart Devices, Environments and Interactions. Wiley, Hoboken (2011)
13. Rincon, J., Poza-Lujan, J.-L., Julian, V., Posadas-Yagüe, J.-L., Carrascosa, C.: Extending MAM5 meta-model and JaCalIV E framework to integrate smart devices from real environments. PloS One **11**(2), e0149665 (2016)
14. Simó-Ten, J.-E., Munera, E., Poza-Lujan, J.-L., Posadas-Yagüe, J.-L., Blanes, F.: CKMultipeer: connecting devices without caring about the network. In: International Symposium on Distributed Computing and Artificial Intelligence, pp. 189–196. Springer (2017)
15. Ström, D.P., Nenci, F., Stachniss, C.: Predictive exploration considering previously mapped environments. In: 2015 IEEE International Conference on Robotics and Automation (ICRA), pp. 2761–2766. IEEE (2015)
16. Tijero, H.P., Gutiérrez, J.J.: Criticality distributed systems through the DDS standard. Revista Iberoamericana de Automática e Informática industrial **15**(4), 439–447 (2018)
17. Yurish, S.Y.: Sensors: smart vs. intelligent. Sens. Transducers **114**(3), I (2010)
18. Zhang, J.: Multi-source remote sensing data fusion: status and trends. Int. J. Image Data Fusion **1**(1), 5–24 (2010)

A Gift-Exchange Model
for the Maintenance of Group Cohesion
in a Telecommunications Scenario

Ana Ramos[1(✉)], Mateus Calado[2], and Luís Antunes[1]

[1] Faculty of Sciences, BioISI–Biosystems and Integrative Sciences Institute,
University of Lisboa, Campo Grande, Lisboa, Portugal
ana.ramos@fc.uan.ao, xarax@ciencias.ulisboa.pt
[2] Departamento de Ciências da Computação, Universidade Agostinho Neto,
Luanda, Angola
padoca@fc.uan.ao

Abstract. In order to extract business information at macro and micro
level and to forecast market trends, companies have a wide range of data
warehouse and business intelligence systems, but the tools to explore or
create scenarios of interactions that customers have with each other are
undervalued. Social simulation is a methodology that can be used in business areas where the understanding of the interactions that customers
have with each other is an added value. In this paper it is presented a
simulation to evaluate the impact of gift-exchange in the cohesion of a
group of people in a telecommunications scenario.

Keywords: Social simulation · Agent Based Models · Gift-exchange ·
Social networks

1 Introduction

The opportunities for application of social simulation in business are numerous,
not only in behavioral economics and artificial markets, but also through the
analysis of factors such as reputation and influence, competition and altruistic
behavior, trust, motivation, among others.

The patterns of consumption in telecommunications emerge fundamentally
from interactions between individuals and Social Simulation has been used to
construct and validate predictive or explanatory models in areas on which the
structure and behavior of the group emerges from simple interactions between
individuals [1].

A concrete scenario for the use of Social Simulation is based on the analysis
of contacts' patterns made between the individuals, to perceive which are the
most dynamic elements of groups.

Another concrete scenario is to verify the contribution of gift-exchange models in group cohesion. Someone's offers obtained through consumption can be

F. Herrera et al. (Eds.): DCAI 2019, AISC 1003, pp. 189–196, 2020.
https://doi.org/10.1007/978-3-030-23887-2_22

given to others, and that may have some impact in terms of maintaining group cohesion. This is the scenario here presented.

The concept and models of gift-exchange have been studied systematically since the mid-twentieth century. Although the interest in its analysis began in Anthropology, other scientific fields as Sociology, Psychology and Economics also ended up studying this theme. Belk [2] points out several functions of gift-exchange, arguing that it has a role of communication, socialization, economic and social exchange. Gift-exchange using Agent Based Models study factors such as influence on reciprocity, cooperation, solidarity and the strengthening of social, economic or political relations. Works as those presented in Alam *et al.* [3], Younger [4] or Rouchier *et al.* [5] are examples of the study of gift-exchange.

Our next topics present the selected methodologies and tools used to develop this work, its design and implementation, the results obtained from the simulation and the conclusions.

2   Methodology and Tools

Agent Based Model (ABM) paradigm has been used since the 1990s in Social Simulation to address problems that are complex, dynamic and decentralized in nature and are difficult to model using a top-down perspective. In [6], Bonabeau presents three advantages of ABM: it captures emerging phenomena, provides a natural description of a system and is flexible.

The ABM paradigm is used because it has a more efficient approach to deal with emerging systems. It models systems as a set of heterogeneous and interdependent agents with the capacity to interact in a decentralized way, in which rules applied are a simplification of reality but complex patterns of behavior can be achieved. In this 'bottom-up' approach, complexity emerges from the interactions between elements (agents), a little like playing with pieces of Lego: pieces are simple, there are not many rules about how to connect them, however very elaborated constructions can be created [7].

To implement the study it was used the agent-based simulation platform provided by NetLogo [8]. NetLogo platform is more oriented to develop and execute simulations (as well as Repast and SeSAm) and also has a high level of popularity among scientific community [9]. It implements its own programming language and has a suitable environment to simulate social and natural phenomena, as mentioned by Tisue and Wilensky in [10].

Railsback *et al.* [11] point out that NetLogo allows the execution of complex simulations. Unlike some other platforms, it is free. This tool is well adapted to the type of environment where agents are created and to the interactions they have, is easy to use, has a clear manual and a library of tested models and also has an interface where it is quick and simple to change the values of simulation's parameters.

3   Design and Implementation

This agent based simulation uses NetLogo platform which provides the bench of experimentation where we created a world populated by agents that interact with each other and with the environment. With it we also created parameters used to test several hypothesis. The purpose of this simulation is to understand how the exchange of gifts contributes to change the cohesion of a group. This group is formed of people who make calls to each other and through incoming calls get available gifts that they can offer to others.

In each simulation the user assigns values to the parameters available on the interface. There is a board in which persons in the group are represented as nodes of a network. The calls and the offers are visually represented as edges between those communicating agents.

3.1   Model Assumptions

Some assumptions were changed or added throughout the development of the work, starting with the simplest ones, which evolved as the model was tested and the consequences that these assumptions had on the model's functioning were verified. The possibility of 'going back' in the simulation and introducing a greater degree of complexity has been mentioned as an advantage of agent-based modeling, for instance by Deichsel and Pyka [12].

In the final version, the model has the following requirements:

– The number of calls in the group is finite: in the real world the time and money to make calls are not unlimited. This is a very strong requirement and one that has remained in the model. However, for the purposes of simplification, calls do not have associated a particular cost or duration.
– People make a defined quantity of calls and this is a parameter of the simulation.
– People who left the group do not call, receive points nor offers.
– People do not have a preferred recipient either to make calls or to offer gifts. This requirement implemented in the model is a simplification of the real.
– Incoming calls are only converted to points when reaching a certain value.
– Points are only converted into gifts when they reach a value stipulated in the simulation.
– People expect to receive a certain number of gifts. If they do not receive them, they lose interest in the group.

3.2   Agents

For the simulation, there are logical units consisting of a person, the particular points, offers and the gifts received (see Fig. 1).

In this way a mapping was created to represent what happens in the real context of the telecommunications, where a person makes calls and can have associated points; in turn the offers and gifts were created to be used in the

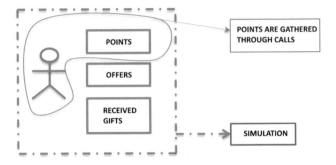

Fig. 1. Logical units in the simulation.

simulation: it is intended to explore how the exchange of offers can contribute to the cohesion of a group. Each logical unit is indivisible and always has all the mentioned elements, even if points, offers and gifts received are equal to zero.

We implemented four different agents. Its main goals are the following:

- 'people', the main goal is to make calls;
- 'point', convert calls in points;
- 'offer', convert points in offers and give them;
- 'received gift', evaluate the maintenance in the group.

4   Simulation's Parameters

To explore the influence of gift exchange on the cohesion of a group, several parameters were created in the simulation. They allow to control the simulation and to diversify the obtained results (see Fig. 2).

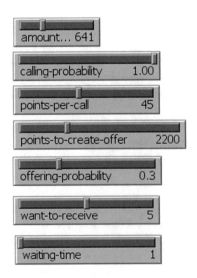

Fig. 2. Sliders available to control simulation's parameters.

- amount: controls the number of people in the simulation.
- points-per-call: indicates how many points each call values.
- points-to-create-offer: controls the amount of points that are necessary to create an offer.
- calling-probability: controls how much probable is to make a call.
- offering-probability: controls how much probable is to give an offer to someone else.
- want-to-receive: defines the minimal amount of gifts that people want.
- waiting-time: time (in steps of the simulation) that agents are available to wait for their gifts.

It was verified that after performing some simulations, it was necessary to fine-tune the parameters values as there were limits that did not produce any changes in simulations' results. This adjustment was an empirical process: simulations were run, results were observed, one parameter was adjusted keeping the others equal, continuing this process until there are useful intervals, which influence the simulation result. Using the Netlogo tool it is easy to make these adjustments, simply changing the intervals of sliders that represent the interface parameters.

5   Simulation Dynamics

In NetLogo platform there were three steps to run the simulation:

- Choose values to the parameters available in the interface;
- Click 'Setup' to create the environment, place the agents in it and initialize variables and the values of parameters;
- Click 'Go' to run the simulation.

The simulation has been constructed so that at each step the agents perform actions that allow them to reach their objectives. In the Fig. 3 are presented the activities and the order at which they happen. Those that belong to the model's initialization (Setup) and those that are part of the execution (Go) are differentiated.

6   Results

In order to explore how the exchange of gifts can influence the maintenance of an element in the group, several scenarios have been tested. Here are four scenarios that represent the results obtained from the executed simulations.

6.1   Scenarios A and B

In these scenarios it was explored the variation of time that people are willing to wait for the gifts. In a world of abundance the effort to obtain points and create offers is the smallest one possible.

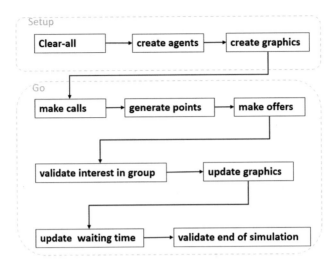

Fig. 3. Activities executed during a simulation.

Scenario A. This type of scenario corresponds to a situation where people have an high expectation of receiving gifts, but are willing to wait for them. The group has an high probability of calling and offering and the cost of obtaining the goods (offers) is low. Here, in the end of the simulation, the amount of people that is inactive (disconnects from the group) does not vary much.

Scenario B. This type of scenario is different from the previous one because although the expectation of receiving gifts is high, the time they are willing to wait has been halved. Here, in the end of the simulation, the number of disconnected people is more variable and the minimum and maximum values of inactive people have increased. There is also a greater uncertainty regarding the results of the simulation.

In both these scenarios abundance makes it easy to keep people in the group, especially when they are willing to wait. When people want to get gifts faster, there is a greater percentage of people disconnecting from the group. The amount of gifts received per person is essentially influenced by the waiting time.

6.2   Scenarios C and D

In scenarios C and D, a world where goods are expensive was simulated. That is, in which to get points you have to receive more calls and to create offers you need to have more points. Interest in receiving gifts is at a maximum value and waiting time varies.

Scenario C. This type of scenario corresponds to a situation where people have a high expectation of receiving gifts and are willing to wait, but the cost of obtaining the goods (offers) is the highest. Here, the minimum and maximum amount of people that remain active does not vary much within each group, they end up almost all inactive. In this scenario of high cost of goods, not only do these values vary little, but it is also quite rare to obtain cases in which all people are not disconnected (about 9/10 of executions ends with everyone disconnected from the group).

Scenario D. In Scenario D the difficulty in getting the points is such that people take too long to get enough points to create offers. The maximum waiting time parameterized in the simulation becomes irrelevant to the total of inactive agents, and thus the two scenarios (C and D) have identical results.

6.3   Conclusions Obtained from Executed Simulations

From the simulations it was concluded the following:

- People's greed and the scarcity of goods are the factors that contribute most to the breakdown of the group.
- If there are many calls and offers, it is possible to maintain a good group cohesion, even when there are high expectations regarding gifts to receive, as long as it is cheap to get points and create offers.
- The cost of goods is irrelevant if people are not interested in receiving gifts, thus mimicking a scenario where people are always in the group, or when people keep the offers for themselves.
- Any scenario is bad for maintaining cohesion if people are not altruistic.

7   Conclusions

The telecommunications service industry is highly competitive. The offerings of various operators are similar, both in terms of price, variability and quality of services. Customer maintenance has several important points, such as the fact that the costs of maintaining the customers are lower for the operators than the ones for having and retaining new customers. The maintenance of customers can contribute to a perception of a better brand image and the attraction of new customers.

From a business perspective, it is not important to implement a somewhat blind gift-exchange model, in which all people earn points with the same effort, because the gifts offered have an associated cost for the company. It is more interesting to use these models if you can identify the 'key' elements of the groups and make the exchange of gifts using those elements to retain other 'satellite' elements. This type of model will have to take into account factors such as the influence of a person in the group and the need that some people have to contact others, even in unfavorable conditions.

Several areas remain open for future research. It can be used data from the real world to test the model, which can be extended using ODD + 2D (ODD +

Decision + Data) methodology explained by Laatabi *et al.* [13]. Another option (it can be implemented simultaneously with the previous one) is to introduce the network of social relations among individuals, trying to identify the most influent actors to the maintenance of the network and thus more interesting from the point of view of customer retention. Changing the business area, but keeping the focus on gift-exchange, models can be created to simulate and test exchanges in environments where there may be an interest or intent to engage in a corruption behavior. The difference between gift giving and bribe can follow the one proposed by Graycar and Jancsics [14], in which they consider that bribes have fundamental differences: at least one of the parts involved are a private or public organization and the exchange is non-transparent.

References

1. Stocker, R., Green, D., Newth, D.: Consensus and cohesion in simulated social networks. J. Artif. Soc. Soc. Simul. **4** (2001)
2. Belk, R.: Gift-giving behavior. College of Commerce and Business Administration, University of Illinois at Urbana, Champaign (1977)
3. Alam, S., Hillebrandt, F., Schillo, M.: Sociological implications of gift exchange in multiagent systems. J. Artif. Soc. Soc. Simul. (2005)
4. Younger, S.: Reciprocity, sanctions, and the development of mutual obligation in Egalitarian societies. J. Artif. Soc. Soc. Simul. (2005)
5. Rouchier, J., O'Connor, M., Bousquet, F.: The creation of a reputation in an artificial society organised by a gift system. J. Artif. Soc. Soc. Simul. (2001)
6. Bonabeau, E.: Agent-based modeling: methods and techniques for simulating human systems. In: PNAS, May, vol. 99, no. suppl. 3 (2002)
7. Calado, M.: Serviço de Emergência Médica Angolano: Optimizaçã Utilizando Sistemas Multi-Agente. Doctoral thesis (2015)
8. NetLogo. http://ccl.northwestern.edu/netlogo/
9. Kravari, K., Bassiliades, N.: A survey of agent platforms. J. Artif. Soc. Soc. Simul. (2015)
10. Tisue, S., Wilensky, U.: NetLogo: design and implementation of a multi-agent modeling environment. In: Proceedings of the Agent 2004 Conference on Social Dynamics: Interaction, Reflexivity and Emergence (2004)
11. Railsback, S., Aylln, D., Berger, U., Grimm, V., Lytinen, S., Sheppard, C., Thiele, J.: Improving execution speed of models implemented in NetLogo. J. Artif. Soc. Soc. Simul. (2017)
12. Deichsel, S., Pyka, A.: A pragmatic reading of Friedman's methodological essay and what it tells us for the discussion of ABMs. J. Artif. Soc. Soc. Simul. (2009)
13. Laatabi, A., Marilleau, N., Nguyen-Huu, T., Hbid, H., Babram, M.A.: ODD+2D: an ODD based protocol for mapping data to empirical ABMs. J. Artif. Soc. Soc. Simul. (2018)
14. Graycar, A., Jancsics, D.: Gift giving and corruption. Int. J. Public Adm. (2016)

New Algorithms

A Sensitivity and Performance Analysis of Word2Vec Applied to Emotion State Classification Using a Deep Neural Architecture

Rodrigo Pasti, Fabrício G. Vilasbôas$^{(\boxtimes)}$, Isabela R. Roque,
and Leandro N. de Castro

Natural Computing and Machine Learning Laboratory (LCoN),
Graduate Program in Electrical Engineering and Computing (PPGEEC),
Mackenzie Presbyterian University, São Paulo, Brazil
rodrigo.pasti@gmail.com, gomesvilasboas@gmail.com,
isabelaruizroqu@gmail.com, lnunes@mackenzie.br

Abstract. Word2Vec has become one of the most relevant neural networks to generate word embeddings for NLP applications. Despite that, little has been investigated in terms of its sensitivity to the word vectors' length (n) and the window size (w). Thus, the present paper performs a sensitivity analysis of Word2Vec when applied to generate word embeddings for a deep neural architecture used to classify emotion states in tweets. Furthermore, we present a computational performance analysis to investigate how the system scales as a function of n and w in different computing environments. The results show that a window size of approximately half the tweet length (8 words) and a value of $n = 50$ suffices to find good performances. Also, by increasing these values one may unnecessarily increase the computational cost.

1 Introduction

Most machine learning algorithms cannot process texts or other types of unstructured data as input [1]. One way of dealing with this problem is by using a distributed representation of words in which vectors preserve the semantic and syntactic meaning of words in a sentence [2,3]. The vectors resulting from this representation are called word embeddings, a concept that has been widely used in Natural Language Processing (NLP) tasks to identify contexts, define relationships between words, perform word translations, extract meaning of texts and many other practical applications [4–6].

Based on this concept, Mikolov et al. [12] introduced the Word2Vec, a predictive model capable of learning word vectors without losing the quality of information for databases with millions of words. Word2Vec has been used to extract contexts and meanings between words from texts and transform into numbers, which is useful as input for a document classification task [7,8]. This is the main avenue of investigation for this paper: to use word vectors as text

© Springer Nature Switzerland AG 2020
F. Herrera et al. (Eds.): DCAI 2019, AISC 1003, pp. 199–206, 2020.
https://doi.org/10.1007/978-3-030-23887-2_23

representations for a deep neural classifier applied to emotion state classification. Also, we assess the Word2Vec sensitivity to its two main input parameters: the word vector dimension (n), and the input window size (w); and investigate its performance when run in an optimized architecture.

The paper is organized as follows. Section 2 provides a brief review of the Word2Vec architecture and the emotion state classification problem, which is the focus of the learning task, and Sect. 2 brings the performance evaluation of Word2Vec applied to this problem, emphasizing its sensitivity to its main input parameters: n and w, and its computational performance when run in different computational architectures. The paper is concluded in Sect. 3 with some general discussions and avenues for future work.

2   Word2Vec and Emotion State Classification

Word2Vec is a feedforward neural network with a single hidden layer: the input layer receives a vector of words in the *one-hot-enconding* format; the hidden layer receives this input vector and calculates the probability distribution of all words in the vocabulary; and the output layer has the same dimension as the input layer. What Word2Vec proposes is that for each word there is a word vector that represents words within the same context as the input word. When presenting each word in the Word2Vec entry it is necessary to define a window (w), which specifies how many words can be considered for the analysis in a sentence, and the number of dimensions (n), which determines how many neurons are used in the hidden layer. The network weights are adjusted using an error back-propagation algorithm, which aims to minimize the error between the network output and the desired output. The Word2Vec weights are initialized with random values. During training, the word vectors are updated using the error back-propagation algorithm and it is necessary to calculate a loss function to update the network weights. Word2Vec uses two supervised algorithms: CBOW and Skip-gram. CBOW predicts words using the past word and the next word in the corpus around a specific target word [9–11]. Skip-gram, by contrast, predicts the surrounding words using the target word [12]. To train and build the corpus, it is necessary to choose between these two algorithms and the corpus will be used later for the classification task.

The sentiment analysis area studies opinions, feelings or emotions, mainly from social media data, about a product, service, company or even a person [13–16]. This area has gained momentum as more and more people express themselves through social media and it becomes possible to classify what they are commenting about some product, company or even a person. This type of classification can be based on polarity, strength or emotions [13,17]. Studies have shown that humans who have difficulties in expressing emotions, cannot maintain and create links easily [16]. In the Psychology literature it is possible to find several theories about what are emotions and what are the basic emotions that a human being can express. The main point is that emotions are crucial to the development of interpersonal relationships [16]. In 1884, William James attempted to define emotion as the change of feeling that occurs when the body perceives an exciting fact [17]. This definition has tried to put an end to the discussion of what emotions are. However, the debate in the scientific community is still vigorous and some scholars have tried not to propose a definition, but to defend a process to define emotions [18]. Ekman and Friessen [19] proposed a theory, largely used on the psychology field, about six basic emotions. They presented some photos from the face of people expressing emotions to an isolated tribe in New Guinea (it was believed that since the tribe is isolated, the emotions they express are pure). The members of the tribe identified six emotions in the photos and the researchers took photos of them representing the same emotions. To conclude the theory, Ekman and Friesen [19] presented the photos of the New Guinea's tribe for other cultures around the world, and all of them were able to identify six emotions: anger; sadness; fear; happiness; disgust and surprise [19]. In the literature it is possible to find some works using robust machine-learning techniques to classify or identify emotions in texts [15, 20–22].

2.1 Sensitivity Analysis of Word2Vec Applied to Classify Emotion States Using a Deep Neural Network

In the experiments to be performed here we used a sample of the dataset presented in [24], which categorizes emotions in seven classes: joy, sadness, anger, love, fear, thankfulness, and surprise. The authors captured five million tweets by monitoring 131 hashtags on emotions between November 10th and December 22nd, 2011. After collection, a series of filtering was performed to improve the quality of their analysis, resulting in a corpus of approximately 2.5 million tweets. In our experiments we used a sample with 35,000 tweets equally divided into the 7 emotion states.

2.2 Deep Neural Net Classifier

In this paper we used a Convolutional Neural Network (CNN) to classify emotion states from tweets. CNNs were used with feedforward layers, allowing the learning of input patterns with a greater amount of aggregated information [25–28, 30], [31]. For example, with vector representations for words and sentences it is possible to represent texts as $n \times m$ dimensional tensors, where n is the dimension of the word vector, and m is the number of word vectors. It is also possible to generate p different arrays of representations, which generates 3D tensors $(n \times m \times p)$ for each sample (tweet).

A CNN with 7 layers was used with the following architecture: a convolutional layer to reduce the dimensionality of the input tensors and to combine word vectors; a Max Pooling Layer, to extract the most relevant features while reducing the dimensionality of the tensors; and a total of 5 feedfoward layers to complete the learning and progressively reduce the dimensionality of the tensors until the network output. The activation functions used in the layers were all hyperbolic tangents.

2.3 Methodology for Emotion State Classification

To investigate the sensitivity of Word2Vec to these two parameters when used to generate the input data for a deep network classifier of emotional states, the following values of n and w were used: $n = 10, 50, 100, 200$ and $w = 2, 4, 8, 12, 16$. The choice of these values was made considering that the literature usually uses Word2Vec to generate word vectors with dimensions between 50 and 200. In the case of w, as tweets have a length limited to 140 characters, the texts usually have less than 32 words and, as the window w is bidirectional, it is sufficient to test with $w \leq 16$. The sensitivity analysis of the two parameters was performed by fixing one of the parameters and varying the other. For the sensitivity analysis to the window size w, a value of $n = 50$ was set for the number of dimensions, and for the sensitivity to the number of dimensions n, $w = 12$. The deep neural network was trained for a fixed number of 300 epochs in all cases, and the performance was measured using the F-measure (F1).

The database was divided into training and validation, and the network with greater generalization capability is stored and its result is considered as the best for a given training set. Bagging was used to generate the data resampling. The training set corresponds to 80% of the whole data and, thus, 20% was used for validation. For each scenario varying the Word2Vector parameters 5 experiments were run, where each one has a variation in the data resampling, as well as the network weight initialization.

2.4 Results and Discussion

Tables 1 and 2 show the performance of the algorithm when varying w and n, respectively. It is possible to observe that by increasing the context window up to the limit, the performance of the algorithm increases, but its improvement becomes marginal for higher values of w. The application of a significance test at a 5% level, indicates that the difference between $w = 12$ and $w = 16$ is not statistically significant. When observing the results in Table 2, it is possible to note that a gain in performance from $n = 10$ to $n = 50$, but a stagnation, and actually a small decrease, in performance for increasing word vector length. Although this may sound counterintuitive, one may reason that longer word vectors are less discriminant than the shorter ones because they allow too many words in the embedding, adding noise words to those that really represent the context.

2.5 Analyzing the Computational Performance

This section presents an analysis of the performance profile of the emotion state classification algorithm in a high-performance processing platform. All experiments were ran on a compute node composed of two Intel® Xeon® Platinum 8160 processors @ 2.10 GHz, each one with 24 physical cores (48 logical) and 33 MB of cache memory, 190 GB of RAM, two Intel® Solid State Drive Data

Table 1. Sensitivity analysis of Word2Vec applied to emotion state classification for varying values of the context window: $w = 2, 4, 8, 12, 16$, and $n = 50$. The results presented are Pr, Re, and F-measure (F1). The results are presented for each of the 7 emotion states separately, and the global average.

w	Measure	Anger	Fear	Joy	Love	Sadness	Surprise	Thankfulness	Global
2	F1	0,710	0,724	0,878	0,978	0,626	0,899	0,976	0,827
	Pr	0,857	0,944	0,896	0,970	0,521	0,906	0,960	0,865
	Re	0,617	0,589	0,869	0,985	0,817	0,898	0,992	0,824
4	F1	0,774	0,803	0,859	0,972	0,698	0,934	0,984	0,861
	Pr	0,901	0,904	0,807	0,959	0,645	0,96	0,982	0,880
	Re	0,689	0,728	0,923	0,986	0,793	0,912	0,987	0,860
8	F1	0,859	0,864	0,923	0,979	0,770	0,954	0,992	0,906
	Pr	0,964	0,975	0,916	0,97	0,659	0,98	0,993	0,922
	Re	0,775	0,776	0,933	0,988	0,927	0,931	0,99	0,903
12	F1	0,886	0,868	0,939	0,984	0,801	0,962	0,993	0,919
	Pr	0,928	0,974	0,93	0,982	0,712	0,987	0,994	0,930
	Re	0,852	0,784	0,949	0,985	0,919	0,938	0,993	0,917
16	F1	0,883	0,905	0,925	0,987	0,819	0,970	0,989	0,925
	Pr	0,952	0,991	0,899	0,985	0,741	0,986	0,983	0,934
	Re	0,824	0,833	0,954	0,989	0,919	0,955	0,994	0,924

Table 2. Sensitivity analysis of Word2Vec applied to emotion state classification for varying values of the word vector length: $n = 10, 50, 100, 200$, and $w = 12$. The results presented are Pr, Re, and F-measure (F1). The results are presented for each of the 7 emotion states separately, and the global average.

n	Measure	Anger	Fear	Joy	Love	Sadness	Surprise	Thankfulness	Global
10	F1	0,848	0,857	0,907	0,942	0,748	0,877	0,985	0,881
	Pr	0,932	0,946	0,936	0,907	0,696	0,825	0,981	0,889
	Re	0,779	0,783	0,881	0,979	0,81	0,938	0,989	0,880
50	F1	0,886	0,868	0,939	0,984	0,801	0,962	0,993	0,919
	Pr	0,928	0,974	0,93	0,982	0,712	0,987	0,994	0,930
	Re	0,852	0,784	0,949	0,985	0,919	0,938	0,993	0,917
100	F1	0,871	0,882	0,947	0,982	0,794	0,968	0,982	0,918
	Pr	0,945	0,981	0,948	0,975	0,697	0,989	0,97	0,929
	Re	0,81	0,802	0,945	0,988	0,924	0,948	0,995	0,916
200	F1	0,869	0,887	0,937	0,982	0,798	0,961	0,987	0,917
	Pr	0,952	0,975	0,929	0,977	0,709	0,976	0,978	0,928
	Re	0,803	0,814	0,948	0,986	0,916	0,947	0,996	0,916

Center (Intel® SSD DC) S3520 Series with 1.2 TB e 240 GB store capacity and a CentOS* 7 operation system running kernel version 3.10.0-693.21.1.3l7.x86_64.

The experiments consisted of running the algorithm 5 times for the same database by varying parameters $n = 10, 50, 100, 200$ and $w = 2, 4, 8, 12, 16$ for the two versions of Python. The final result was obtained by taking the mean of the total execution time for the 5 runs for each configuration in each Python version. The execution with 96 threads was defined for the Word2Vec and TensorFlow modules. The neural network will perform two experiments with 50 epochs each, because the number of experiments and the number of epochs influence the accuracy of the model and the goal here is to analyze the execution time profile.

Figure 1 shows the average running time for the different values of n and w. It is possible to observe that when we increase the value of n, there is a significant increase in the execution time in both environments. This is because n defines the number of dimensions of the neural network input tensors. Therefore, by increasing its value, we are increasing the complexity of the algorithm. The same behavior was not observed when we increased the value of w. This is because its variation was not enough to increase the complexity of the operations that evolve.

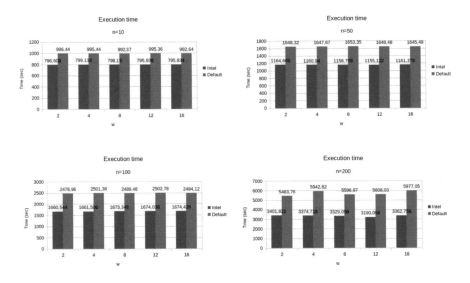

Fig. 1. Running time for the different values of n and w.

When we compare the execution time between the environments, we see that the Intel environment presented a shorter execution time in all cases. This indicates that the greater the complexity of the algorithm, the greater the disparity between the execution times of the analyzed environments.

3   Discussion and Future Works

This paper aimed to investigate the influence of the main adjustable parameters of Word2Vector when applied in the classification of emotional states using a deep neural architecture. The word vectors dimensions directly impact the cardinality of the generated tensor. The greater the number of dimensions, the greater the computational effort to train the network, but not always a better learning is observed. The window w between words has a direct impact on the definition of context and vectors' spreading in the Euclidean space, where a larger number implies a greater number of neighboring words. It was observed that after some value of w the gain in performance is not significant. The main issue to be investigated as future research is if these conclusions can be generalized for other types of text data.

Acknowledgements. The authors thank CAPES, CNPq, Fapesp, and Mackpesquisa for the financial support. The authors also acknowledge the support of Intel for the Natural Computing and Machine Learning Laboratory as an Intel Center of Excellence in Artificial Intelligence.

References

1. De Castro, L.N., Ferrari, D.G.: Introduction to Data Mining: Basic Concepts, Algorithms, and Applications. Saraiva (2016). (in Portuguese)
2. Hinton, G.E.: Learning distributed representations of concepts. In: Proceedings of the Eight Annual Conference of the Cognitive Science Society (1986)
3. Rumelhart, D.E., Hinton, G.E., Williams, R.J.: Learning representations by back-propagation errors. Nature **323**, 533–536 (1986)
4. Tang, D., Wei, F., Yang, N., Zhou, M., Liu, T., Qin, B.: Learning sentiment-specific word embedding for Twitter sentiment classification. In: 52nd Annual Meeting of the Association for Computational Linguistics, Baltimore, Maryland, USA (2014)
5. Zhou, G., He, T., Zhao, J., Hu, P.: Learning continuous word embedding with metadata for question retrieval in community question answering. In: 53nd Annual Meeting of the Association for Computational Linguistics and the 7th International Joint Conference on Natural Language Processing, Beijing, China (2015)
6. Wang, P., Xu, B., Xu, J., Tian, G., Liu, C.-L., Hao, H.: Semantic expansion using word embedding clustering and convolutional neural network for improving short text classification. Neurocomputing **174**, 806–814 (2016)
7. Dos Santos, C., Gatti, M.: Deep convolutional neural networks for sentiment analysis of short texts. In: International Conference on Computational Linguistics (2014)
8. Koper, M., Kim, E., Klinger, R.: Emotion intensity prediction with affective norms, automatically extended resources and deep learning. In: Proceedings of the 8th Workshop on Computational Approaches in Subjectivy Sentiment and Social Media Analysis, pp. 50–57 (2017)
9. Goldberg, Y.: A primer on neural network models for natural language processing. J. Artif. Intell. Res. **57**(2016), 345–420 (2016)
10. Sharma, R., Kaushik, P.: Literature survey of statistical, deep and reinforcement learning in natural language processing. In: 2017 International Conference on Computing, Communication and Automation (ICCCA) (2017)
11. Mulder, W.D., Bethard, S., Moens, M.-F.: A survey on the application of recurrent neural networks to statistical language modeling. Comput. Speech Lang. **30**(1), 61–98 (2015)
12. Mikolov, T., Chen, K., Corrado, G., Dean, J.: Efficient estimation of word representations in vector space. In: Proceedings of Workshop at ICLR (2013)

13. Liu, B.: Sentiment analysis and opinion mining. Synth. Lect. Hum. Lang. Technol. **5**, 1–167 (2012). Morgan & Claypool Publishers
14. Pang, B., Lee, L.: Opinion mining and sentiment analysis. Found. Trends® Inf. Retrieval **2**, 1–135 (2008)
15. Bucar, J., Povh, J.: Sentiment analysis in web text: an overview. In: Recent Advances in Information Science, pp. 154–159 (2013)
16. Thelwall, M., Buckley, K., Paltoglou, G., Cai, D.: Sentiment strength detection for the social Web. J. Am. Soc. Inf. Sci. Technol., 2544–2558 (2010)
17. Lima, A.C.E.S., de Castro, L.N.: Automatic sentiment analysis of Twitter messages. In: Proceedings of the Fourth International Conference on Computational Aspects of Social Networks (2012)
18. Weiyuan, L., Hua, X.: Text-based emotion classification using emotion cause extraction. Expert Syst. Appl. **41**(4), 1742–1749 (2014)
19. Ekman, P., Friesen, W.V.: Constants across cultures in the face and emotion. J. Pers. Soc. Psychol. **17**, 124–129 (1971)
20. Ekman, P., Friesen, W.V., Ellsworth, P.: Emotion in the Human Face, 1 edn., vol. 1 (1972). (A. P. Goldstein and L. Krasner, Eds., Pergamon)
21. Bellegarda, J.R.: Emotion analysis using latent affective folding and embedding. In: Proceedings of the NAACL HLT 2010 Workshop on Computational Approaches to Analysis and Generation of Emotion in Text, Los Angeles, California (2010)
22. Chaffar, S., Inkpen, D.: Using a heterogeneous dataset for emotion analysis in text. In: Proceedings of the 24th Canadian Conference on Advances in Artificial Intelligence, St. John's, Canada (2011)
23. Dosciatti, M.M., Paterno, L., Paraiso, E.C.: Identificando Emoções em Textos em Português do Brasil usando Máquina de Vetores de Suporte em Solução Multiclasse," Encontro Nacional de Inteligência Artificial e Computacional (ENIAC) (2013)
24. Wang, W., Chen, L., Thirunarayan, K., Sheth, A.P.: Harnessing Twitter 'Big Data' for automatic emoticon identification. In: 2012 International Conference on Social Computing, 11 January (2013)
25. Jurafsky, D., Martin, J.H.: Speech and Language Processing, 2nd edn. Prentice Hall, Upper Saddle River (2009)
26. Kalchbrenner, N., Grefenstette, E., Blunsom, P.: A convolutional neural network for modelling sentences. In: Proceedings of the 52nd Annual Meeting of the Association for Computational Linguistics (2014)
27. Lai, S., Xu, L., Liu, K., Zhao, J.: Recurrent convolutional neural networks for text classification. In: AAAI 2015 Proceedings of the Twenty-Ninth AAAI Conference on Artificial Intelligence (2015)
28. Schmidhuber, J.: Deep learning in neural networks: an overview. Neural Netw. **61**, 85–117 (2015)
29. Da Silva, I.R.R., Lima, A.C.E.S., Pasti, R., De Castro, L.N.: Classifying emotions in twitter messages using a deep neural network. Springer (2018)
30. Da Silva, I.R.R., De Castro, L.N.: Estudos sobre um modelo de representação distribuída de palavras no contexto de análise de estados emocionais. Master Thesis (Master in Electrical and Computer Engineering) – Mackenzie Presbiteryan University, Sao Paulo (2018)

Multi-view Cooperative Deep Convolutional Network for Facial Recognition with Small Samples Learning

Amani Alfakih[⊠], Shuyuan Yang, and Tao Hu

Department of Electrical Engineering, Xidian University, Xi'an 710071, China
am775901039@gmail.com, syyang2009@gmail.com,
1010004295@qq.com

Abstract. In the community of computer vision, deep learning has been widely applied in many classification tasks. However, the performance of deep networks depends heavily on the large number of labeled samples. In this paper, we propose a multi-view Deep Convolutional Neural Network to recognize facial expression while very small number of samples is available. First, facial images are downsampled to different scales and upsampled as multi-view samples. Then a multi-view DCNN is constructed with twin structure and cooperative learning. After one channel is trained by single view samples, the parameter is transferred to another channel for fine tuning using another view samples. Some experiments are taken on FER2013 and RAF datasets, and the experimental results illustrate that the proposed multi-view DCNN network has a good performance where achieves 72.27% on the private set of FER2013 dataset, and the transfer DCNN model achieves 83.08% on the test set of RAF database.

Keywords: Facial expression recognition · Multi-view · Convolutional neural networks · Transfer learning

1 Introduction

Facial Expression Recognition is a hot topic in the fields of artificial intelligence and computer vision, a lot of works and research have been published in this field, where there are many reasons that make it still a challenging such as variations in background, expression, lighting and position etc. The techniques of computer vision and machine learning to recognize facial expressions find their ways of designing a new generation of human computer interfaces. Ekman et al. identified the six facial expressions (happiness, sadness, disgust, fear, angry and surprise) as basic facial expressions that are universal among human beings [1]. A new method was presented in face recognition which used Spares Representation and combined it with Least-Squares (SR +RLS) all details in [2]. Best-Rowden and Jain investigated how the recognition performance is affected by facial aging, especially when we have a large population, Multilevel statistical models were used to genuine similarity score, especially with regard to increase the elapsed time between two face images [3]. A simple technique using Morphological and Histogram method was proposed for classification of facial expression, and it only deals with the static images [4], also HOGs features were used

© Springer Nature Switzerland AG 2020
F. Herrera et al. (Eds.): DCAI 2019, AISC 1003, pp. 207–216, 2020.
https://doi.org/10.1007/978-3-030-23887-2_24

for face recognition in [5]. Perikos et al. introduced classification schema on two stages: using a Support Vector Machine (SVM) to recognize whether the expression is emotional or is neutral and then a Multilayer Perceptron Neural Network (MLPNN) was used to specify each expression's emotional content [6].

A lot of success in a wide variety of image classification tasks when Convolutional Neural Network (CNN) has been used and achieved good performance [7]. Some investigations were presented about the reason why we get results much better for classification when the hand crafted features extraction are replaced by CNNs [8], the authors used the fusion of multiple CNNs and metric learning. Zhou et al. proposed a new method which is called multi-scale CNNs for facial expression recognition, their method consisted of three sub CNNs with different input images size [9]. A deep neural network architecture was proposed to process and explain the facial expression recognition (FER) problem using multiple well-known standard face datasets. This network contained two convolutional layers, max pooling layer after each convolutional layer and then four inception layers [10]. A novel approach was applied to represent the face using a set of Convolutional Neural Networks designed to extract features from face images followed by three layer Stacked Auto Encoder (SAE) to compress the concatenated dimensions [11].

The work in [12], authors suggested the feature selection network (FSN) to get accurate facial details effectively and disregard the negative impact of background, feature selection mechanism was inserted within AlexNet before the fully connected layer. Giannopoulos et al. studied deep learning performance methods to recognize facial expressions. Specifically, to recognize presence of emotional content and then to recognize the precise emotional content of the expressions of face. Their experimental study dealt two deep learning approaches performances that are GoogLenet and Alexnet [13]. The authors in [14], studied Dense SIFT and regular SIFT and comparison when combined with CNN features for facial expression recognition. Yosinski et al. proposed a new technique to improve the performance of deep neural network, their technique had measured transferability of the features from any layer of the neural network, which uncovers their general or specific features [15].

In this work, we illustrate that the DCNN approaches can be effectively when we apply a multi-view DCNN model and the performance has improved. Furthermore, deep models on two small datasets for recognizing facial expression are applied. Thus, to acquire efficient result, the transfer learning is applied.

This paper is structured as follows. Section 2 explains the proposed deep Convolutional neural network and multi-view DCNN network to recognize facial expression. Experiments on facial expression recognition and discussion about the performance are presented in Sect. 3. Finally, in Sect. 4, our conclusions are drawn.

2   Methodology

2.1   DCNN

Facial expression recognition is being useful in many systems and applications. In order to understand the condition of humanity in a deeper way, we need to give the ability and knowledge to computer applications to be able to recognize the emotional state of human facial expressions; therefore, the task itself is very challenging in the presence of many applications. This section illustrates the proposed network for recognition facial expressions, which is based on CNN. We create a deep CNN architecture and designed it from scratch. Our proposed DCNN model consists of eight convolutional layers followed by two fully connected layers. We apply batch normalization after each convolutional layer then Rectified Linear Unit is used as an activation function. Then, we apply a dropout value of 25% of neurons between each two layers in our model and after the last fully connected layer, it is used to reduce the overfitting of network. The output of the last dense layer is fed to softmax with seven ways, which represent the facial expressions at the network last stage. The softmax is represented by,

$$y_i = \text{softmax}(x_i) = \frac{\exp(x_i)}{\sum_{j=1}^{N}(\exp(x_j))} \tag{1}$$

where x_i is the input sample, N the number of classes and y_i the prediction score to belong to the ith class.

The number of kernels of the first and second convolutional layer is 64 kernels, from third to sixth layer is 128, and the last two convolutional layers is 512. Our model has used the same size of kernel at each layer is 3×3, except the third layer is 5×5. Max-pooling layer is used with the stride is 2 before we apply dropout in third layer and from sixth to eighth layers; it reduced the deeper layers computation. The output of the last convolution layer is fed to the first dense layer, after it is flattened. The first dense layer consists of 256 neurons which are linked as the fully connected layer, while the second layer consists of 512 neurons.

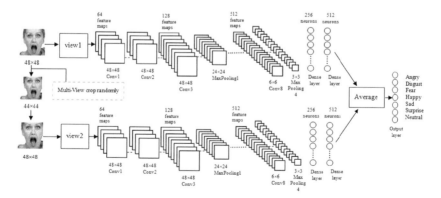

Fig. 1. Structure of the multi-view DCNN method

2.2   Multi-view DCNN Method

We work to improve the performance of facial expression recognition by combining multi-view learning, deep convolutional neural network and cooperative learning. The idea of cooperative learning is the networks work together in order to improve each other's learning. In this approach, we applied cooperative learning using two networks which work together to enhance the results. We proposed a multi-view DCNN with twin structures of DCNNs which are trained, separately. First network is using the original FER2013 dataset and the second is using the reconstructed images for training, all details about the structure of multi-view DCNN method are shown in Fig. 1. The reconstructed images are gotten by cropping randomly the original images to different scales and resizing them as multi-view samples. Cooperative learning is applied by combining the pretrained models that seek to attained the best performance by removing the output layer from both models and take the features from last fully connected layers to combine them. Then, the outputs of average become the input to the last layer which is softmax layer for seven classes for an image.

3   Experiments

3.1   Datasets

We conduct experiments on the FER2013 [18] and RAF [17] datasets. The Facial Expression Recognition 2013 database was published by Pierre Luc Carrier and Aaron Couville in the ICML 2013 Challenges in Representation Learning. FER2013 database contains 35,887 facial images most of them in wild settings and the data are in form grayscale. The training set consists of 28,709 samples, and both the public (validation) test and private test sets have 3,589 samples. All the images in FER2013 dataset were categorized in to one of seven categories: fear, disgust, angry, happy, sad, surprise and neutral. RAF-DB is a large database of facial expression images was collected directly from the Internet. The database contains 15,339 real-world images, which were split into two parts a training set with 12,271 samples and 3,068 samples as a test set. RAF has six basic emotional categories (surprise, fear, disgust, angry, happy, sad) and neutral. 315 human coders categorized the images in RAF, where the crowdsourcing techniques were used to determine the final annotations. Figure 2 shows some examples of facial images from FER2013 and RAF datasets. It is obvious in Fig. 2, the images in both datasets have more complex background, characters, pose, angle, illumination etc. then compared to most existing facial databases.

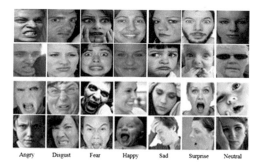

Angry   Disgust   Fear   Happy   Sad   Surprise   Neutral

Fig. 2. Examples of facial images of FER2013 and RAF datasets. The colored images belong to RAF and the gray images belong to FER2013

3.2 Implementation Details

In FER2013 dataset, the size of input images is 48 × 48 pixels. We don't need to apply alignment of face due to the images are small size, each image is normalized to have zero mean and unit variance. In term of RAF dataset, the input images of 100 × 100 pixels are downscaled to 48 × 48 pixels for DCNN and transferred model, all images are converted to grayscale. In both datasets, to augment the number of samples, data augmentation is applied; we apply different linear transformations which include shifting randomly images horizontally and vertically, horizontal flipping and rotation. We use Python software to implement our idea [16], where a Deep Convolutional Neural Network is created successfully using Keras which is the neural network library.

Our work consists of three steps. Firstly, we train our DCNN network from scratch on 28709 samples of the FER2013 dataset. Secondly, the same network is used to train the same data but after random downsampled to 44 × 44 and upsampled to 48 × 48. During training all networks, the hyper-parameters are tuned by using the public set as validation set, while we use the private set as test set. The weights are initialized randomly, and our DCNN is trained for 200 iterations, the second model for 500 iterations, both of them with batch size is 128. In addition, we conduct training on RAF dataset, we selected randomly 10% from training samples to use as validation set. Thirdly, we train DCNN model second time from scratch on 11,043 samples of the RAF database, we got a good performance after the training continues 200 iterations. Also, we use weights of our pretrained model which is trained on FER2013 as initial weights for the training on RAF database, our model is trained for 100 iterations in this state. The cost function which is used in our research is a categorical crossentropy and it is optimized using Adam.

3.3 Results on FER2013

Figure 3 Provides the results of our models on the validation and test sets of FER2013 dataset. The proposed DCNN model achieves an accuracy of 67.52% on the test set and

65.36% on validation set after it was trained 200 iterations. When we use data augmentation the accuracy is increased to 71.6% and 70.37% on test and validation sets,
respectively. The convergence curves between training accuracy and validation accuracy using 200 iterations and with data augmentation are shown in Fig. 4(a). Our model
performs better when we apply multi-view DCNN method, which achieves accuracy up
to 70.55% and 72.27% on public and private test sets, respectively. Figure 5 shows the
normalized confusion matrices of DCNN, DCNN+ data_Aug and multi-view DCNN
on the FER2013 test set.

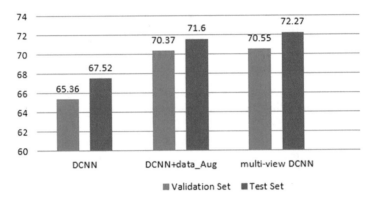

Fig. 3. The performance of our models on FER2013 dataset.

3.4   Results on RAF

DCNN with data augmentation achieves 81.7% on the test set of RAF dataset after
training 200 iterations. When we use the pretrained model which is trained on
FER2013 to fine-tune on RAF dataset, the accuracy improved to 83.08%. Figure 4(b)
shows the convergence curves during the training process of DCNN model on RAF
dataset. Confusion matrices of our models on RAF are shown in Fig. 6.

We compare our results with some earlier techniques which were applied the
testing on FER2013 and RAF datasets. The accuracy results of these methods and our
method are shown Tables 1 and 2. Obviously, from Table 1. The multi-view DCNN
model performs 0.47%, 1.11% higher accuracy compared with Multi-Scale CNN and
winner method, respectively. According to Table 2. Transfer DCNN model performs
12.1%, 9.89% higher accuracy compared with DLP-CNN(LDA) and Boosting-POOF,
respectively. Overall, the performance of our methods is better than others. Note that,
the best results are highlighted in bold.

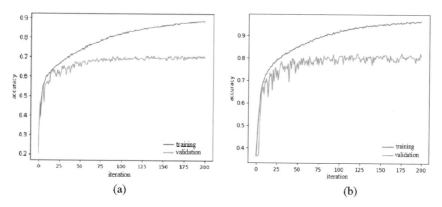

Fig. 4. The convergence curves using the proposed DCNN + data_Aug. (a) On FER2013. (b) On RAF.

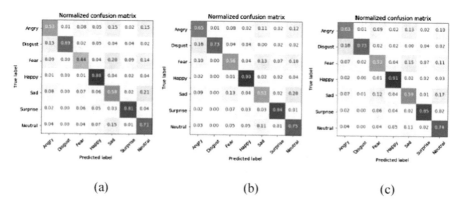

(a)   (b)   (c)

Fig. 5. Confusion matrices of proposed models on FER2013 test set. (a) Using DCNN without data augmentation. (b) Using DCNN+ data_Aug. (c) Using multi-view DCNN.

Table 1. Comparative the recognition accuracy among the proposed methods and some state-of-the-art methods on FER2013.

Method	Validation set acc	Test set acc
The winner of FER2013 [19]	69.77%	71.16%
Multiple Deep Network Learning [21]	70.1%	72.1%
Multi-Scale CNN [9]	69.82%	71.8%
DCNN(our)	70.37%	71.6%
Multi-view DCNN(our)	**70.55%**	**72.27%**

Table 2. Comparative the recognition accuracy among the proposed methods and some state-of-the-art methods on RAF.

Method	RAF acc.
VGG	58.22%
DLP-CNN(LDA) [17]	70.98%
FSN [12]	72.46%
Boosting-POOF [20]	73.19
DCNN(our)	81.7%
Transfer DCNN(our)	**83.08%**

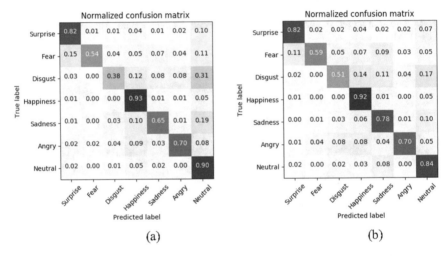

Fig. 6. Confusion matrices of proposed models on RAF database test set. (a) Using proposed DCNN+ data_Aug. (b) Transfer DCNN.

4   Conclusion

In this work, we have presented deep models to recognize facial expressions on based on deep convolutional neural network. Multi-view DCNN method is implemented to improve the performance of our network, where the accuracy reaches 72.27% on FER2013 dataset. As saw as data augmentation greatly affects the performance, where the accuracy is increased significantly when we apply it with our models. Moreover, the results have improved considerably on RAF database, it achieves 83.08% by using transfer DCNN. The experimental results show that our DCNN has good generalization ability. In future work, we aim to combine our model with Deep learning approaches and evaluate them on different datasets.

References

1. Ekman, P., Friesen, W.V.: Constants across cultures in the face and emotion. J. Pers. Soc. Psychol. **17**(2), 124–129 (1971)
2. Iliadis, M., et al.: Sparse representation and least squares-based classification in face recognition. In: 2014 22nd European Signal Processing Conference (EUSIPCO), pp. 526–530. IEEE Xplore (2014)
3. Best-Rowden, L., Jain, A.K.: A longitudinal study of automatic face recognition. In: International Conference on Biometrics (ICB) 2015, pp. 214–221. IEEE Xplore (2015)
4. Aung, D.M., Aye, N.: A Facial expression classification using histogram based method. In: 2012 4th International Conference on Signal Processing Systems, vol. 58, pp. 1–5 (2012)
5. Déniz, O., et al.: Face recognition using histograms of oriented gradients. Pattern Recogn. Lett. **32**(12), 1598–1603 (2011)
6. Perikos, I., Ziakopoulos, E., Hatzilygeroudis, I.: Recognize emotions from facial expressions using a SVM and neural network schema. In: Engineering Applications of Neural Networks, pp. 265–274. Springer, Heidelberg (2015)
7. Sun, Y., Wang, X., Tang, X.: Deep convolutional network cascade for facial point detection. In: 2013 Conference on Computer Vision and Pattern Recognition (CVPR), pp. 3476–3481. IEEE (2013)
8. Hu, G., et al.: When face recognition meets with deep learning: an evaluation of convolutional neural networks for face recognition. In: 2015 IEEE International Conference on Computer Vision Workshop (ICCVW), pp. 384–392 (2015)
9. Zhou, S., et al.: Facial expression recognition based on multi-scale CNNs. In: Chinese Conference on Biometric Recognition, vol. 9967, pp. 503–510. Springer, Heidelberg (2016)
10. Mollahosseini, A., Chan, D., Mahoor, M.H.: Going deeper in facial expression recognition using deep neural networks. In: IEEE Winter Conference on Applications of Computer Vision (WACV), pp. 1–10. IEEE (2016)
11. Ding, C., Tao, D.: Robust face recognition via multimodal deep face representation. IEEE Trans. Multimedia **17**(11), 2049–2058 (2015)
12. Zhao, S., Cai, H., Liu, H., Zhang, J., Chen, S.: Feature selection mechanism in CNNs for facial expression recognition. In: 2018 British Machine Vision Conference (BMVC), pp. 1–12 (2018)
13. Giannopoulos, P., Isidoros P., Hatzilygeroudis, I.: Deep learning approaches for facial emotion recognition: a case study on FER-2013. In: Advances in Hybridization of Intelligent Methods, vol. 85, pp. 1–16. Springer, Heidelberg (2018)
14. Al-Shabi, M., Cheah, W.P., Connie, T.: Facial expression recognition using a hybrid CNN-SIFT aggregator. In: 2017 International Workshop on Multi-disciplinary Trends in Artificial Intelligence, vol. 10607, pp. 139–149. Springer, Heidelberg (2017)
15. Yosinski, J., Clune, J., Bengio, Y., Lipson, H.: How transferable are features in deep neural networks? In: 27th International Conference (NIPS), vol. 2, pp. 3320–3328 (2014)
16. Sanner, M.F.: Python: a programming language for software integration and development. J. Mol. Graph. Model. **17**(1), 57–61 (1999)
17. Li, S., Deng, W., Du, J.: Reliable crowdsourcing and deep locality-preserving learning for expression recognition in the wild. In: 2017 IEEE Conference on Computer Vision and Pattern Recognition (CVPR), pp. 2584–2593. IEEE (2017)
18. Goodfellow, I., Erhan, D., Luc Carrier, P., et al.: Challenges in representation learning: a report on three machine learning contests. Neural Netw. **1**(64), 59–63 (2015)

19. Tang, Y.: Deep learning using linear support vector machines. In: 2013 International Conference on Machine Learning, vol. 28 (2013)
20. Liu, Z., Li, S., Deng, W.: Boosting-POOF: boosting part based one vs one feature for facial expression recognition in the wild. In: 2017 IEEE International Conference on Automatic Face & Gesture Recognition, pp. 967–972. IEEE (2017)
21. Yu, Z., Zhang, C.: Image based static facial expression recognition with multiple deep network learning. In: 2015 Proceedings of the ACM on International Conference on Multi-modal Interaction, pp. 435–442 (2015)

Sampling Rank Correlated Subgroups

Mohamed-Ali Hammal[1], Bernardo Abreu[2], Marc Plantevit[3],
and Céline Robardet[1(✉)]

[1] Université de Lyon, CNRS, INSA Lyon, LIRIS, UMR5205, 69621 Lyon, France
{mohamed-ali.hammal,celine.robardet}@insa-lyon.fr
[2] Universidade Federal de Minas Gerais (UFMG), Belo Horizonte, Brazil
bernardoabreu@dcc.ufmg.br
[3] Université de Lyon, CNRS, Université Lyon 1, LIRIS UMR5205,
69622 Lyon, France

Abstract. Data mining, a key technique in knowledge discovery, is the process of identifying useful patterns from a collection of data. This process is made difficult for complex data combining, for example, numeric and symbolic attributes, or also when the number of observations is large. In this paper, we present a pattern mining approach to identify local correlations in the data, that is to say, sets of numerical attributes that strongly co-vary together in a subset of the data. The sets of numerical attributes and the subset of data are automatically (inductively) identified by the method. Whereas the space of patterns to be potentially explored is exponential, the complexity of the problem can be overcome by using sampling techniques that have several advantages: (1) reducing the computation cost, (2) identifying most important patterns, and (3) making possible to process large databases by distributed computing on multiple machines.

Keywords: Data mining · Markov Chain sampling ·
Correlated subgroups

1 Introduction

Data mining is a knowledge discovery technique that automatically selects and aggregates information into patterns and correlations. It can be used to formulate various hypotheses using nontrivial processes that identify "valid, novel, potentially useful, and ultimately understandable" [1] structures in the data. The main difficulties encountered by data mining techniques are related to the complexity of the data that makes it difficult the identification of patterns of interest and also affects the performance of the data mining process adversely. The complexity comes from the fact the data is generally noisy and incomplete, but also from the heterogeneity of the type of its attributes: algorithms may handle numeric and symbolic descriptors, while maintaining a high level of performance.

© Springer Nature Switzerland AG 2020
F. Herrera et al. (Eds.): DCAI 2019, AISC 1003, pp. 217–225, 2020.
https://doi.org/10.1007/978-3-030-23887-2_25

In this paper, we consider the challenge of identifying interesting patterns to depict observations described by numeric and symbolic attributes. We propose to look for correlations between numeric attributes to pinpoint possible relationships between them. Such correlations are all the more interesting when they occur locally in the dataset: while the age and income variables are not globally correlated, they become so when restricting the observations to the managerial employees. The pattern mining approach we propose identifies patterns that are locally correlated in the data: sets of numerical attributes (e.g. age and income) that strongly co-vary together in a subset of the data, identified by restrictions on the values of some others attributes (e.g. category of employee = manager). The sets of numerical attributes and the subset of data are automatically (inductively) identified by the method and patterns with high value on an interestingness measure that evaluates its local correlation are retrieved.

Whereas the space of patterns to be potentially explored is exponential, the complexity of the problem can be overcome by using sampling techniques that make it possible to reduce the computation cost while identifying most important patterns. In what follows, we formally define rank correlated subgroups and present related work on the sampling algorithms designed for pattern mining problems. Then, we detail the sampling technique we use and show that it makes possible to obtain high quality patterns in an efficient way, by comparing the obtained results with the ones computed by an exhaustive approach. This sampling approach computes each pattern independently from the others and thus enables to process large databases by distributing the computation over multiple machines.

2   Rank Correlated Subgroups

Suppose we have a set of observations $\mathcal{O}$ described by a set of numerical attributes $\mathcal{C}$. We can obtain the correlation between two of such attributes with the Kendall τ correlation measure that evaluates whether the observations have a similar rank when ordered by each of the two variables. Numerically, it is proportional to the number of pairs of observations ordered in the same way on each of the attributes:

$$\tau(ab) = \frac{|\eta(ab)| - |\overline{\eta(ab)}|}{\frac{1}{2}n(n-1)}, \ a, b \in \mathcal{C}$$

with $\eta(ab) = \{(o_i, o_j) \in \mathcal{O}^2 \mid ((a(o_i) < a(o_j))$ and $(b(o_i) < b(o_j)))$ or $((a(o_i) > a(o_j))$ and $(b(o_i) > b(o_j)))\}$, the number of concordant pairs of observations, $\overline{\eta(ab)} = \mathcal{O}^2 \setminus \eta(ab)$, the set of discordant pairs, and n the number of observations.

We can generalize this measure for more than 2 attributes by adding a sign ($\{+, -\}$) to each attribute of the pattern. By convention, the sign of the first attribute is set to $+$ to filter out symmetric patterns that depict the same piece of information (e.g., $a^+ b^+ c^+ \equiv a^- b^- c^-$).

Definition 1 (Rank-Correlation pattern). *A rank-correlation pattern C is a set of (at least two) signed attributes from $\mathcal{C}$ defined as $C = \{(a, s) \mid a \in \mathcal{C}$ and $s \in \{-, +\}\}$. The sign of the first attribute in the canonical order is $+$ by convention. Given a set of observations $O \subseteq \mathcal{O}$, the set of concordant pairs from O with pattern C is given by $\eta(C, O) = \{(o_i, o_j) \in O \times O \mid \nu_C(o_i, o_j)\}$ with*

$$\nu_C(o_i, o_j) \equiv \bigwedge_{(a,s) \in C} (a(o_i) <_s a(o_j))$$

and $<_s$ is the conventional binary relation on $\mathbb{R}$ $<$ when $s = +$, and $>$ when $s = -$. The Kendall's τ rank correlation measure of C on O is then:

$$\mathcal{T}(C, O) = \frac{|\eta(C, O)|}{N(O)} \text{ with } N(O) = \binom{|O|}{2}. \tag{1}$$

Evaluating the correlations between numerical attributes across all observations is of little use because this global information about the dataset is generally known to the end user. What has more added values is to identify a part of the observations where the correlation is abnormally high. Such observations, that form a subgroup, are defined by means of a conjunction of conditions on their values on some descriptive attributes $\mathcal{R}$. The set $\mathcal{R} = \langle d_1, \ldots, d_{|\mathcal{R}|} \rangle$ is made of numeric and symbolic attributes, and the conditions can be seen as restrictions on their value domains. For example, let us consider three attributes: AGE with $\mathbf{Dom}(Age) = [25, 60]$, GENDER with $\mathbf{Dom}(Gender) = \{female, male\}$ and INCOME with $\mathbf{Dom}(Income) = [1000, 10000]$. A subgroup can be defined as $Age \in [30, 40]$ and $Gender \in \{female\}$ and $Income \in [2500, 4000]$.

Definition 2 (Subgroup and support). *A subgroup of observations is given by a description $D = \langle f_1, \ldots, f_{|\mathcal{R}|} \rangle$ where each f_ℓ is a restriction on the domain of the attribute $d_\ell \in \mathcal{R}$:*

- *If d_ℓ is nominal, then $f_\ell = \{v\}$ with $v \in \mathbf{Dom}(d_\ell)$, or $f_\ell = \mathbf{Dom}(d_\ell)$*
- *If d_ℓ is numerical, then $f_\ell = [v, w]$ with $v, w \in \mathbf{Dom}(d_\ell)$ and $v < w$*

The set of observations of $\mathcal{O}$ that satisfy D is:

$$\sigma(D) = \{o_i \in \mathcal{O} \mid d_\ell(o_i) \in f_\ell, \forall \ell = 1 \ldots |\mathcal{R}|\}$$

We say that D is frequent if its support $supp(D) = |\sigma(D)| \geq \alpha$.

Definition 3 (Correlated subgroup). *A pair (C, D), with C a rank-correlation pattern and D a frequent subgroup, is correlated if $\mathcal{T}(C, \sigma(D)) \geq \beta$.*

The unexpectedness of a pattern can be measured with the Weighted Relative Accuracy (WRAcc) [2] that evaluates the difference between the local and the global Tau measures weighted by the proportion of the subgroup in $\mathcal{O}$.

Definition 4 (WRAcc). *The WRAcc measure of (C, D) is defined as:*

$$\mathbf{WRAcc}(C, D) = \frac{N(supp(D))}{N(\mathcal{O})} (\mathcal{T}(C, \sigma(D)) - \mathcal{T}(C, \mathcal{O})) \tag{2}$$

Some correlated subgroups can be considered as equivalent as they share the same support. Among them, the most interesting patterns are the maximal ones: the frequent correlated subgroups $(\sigma(D) \geq \alpha)$ whose supersets are not.

Definition 5 (Maximal correlated subgroups). *Let (C_1, D_1) and (C_2, D_2) two correlated subgroups. We say that (C_2, D_2) is more specific than $(C_1, D1)$, denoted $(C_1, D_1) \preceq (C_2, D_2)$ if $\forall (a, s) \in C_1$, $(a, s) \in C_2$ and $\forall f_\ell \in D_1$, $f_\ell \subseteq g_\ell$, with $g_\ell \in D_2$. (C_2, D_2) is maximal iff $\forall (C_1, D_1)$ such that $supp(D_1) \geq \alpha$, $(C_1, D_1) \not\preceq (C_2, D_2)$.*

Our mining task can be expressed as the *maximal correlated subgroup mining problem* that consists in computing the collection $\mathcal{M}$ of maximal correlated subgroups defined as $\forall (C, D) \in \mathcal{M}$, (C, D) is maximal, $supp(D) \geq \alpha$, $\tau(C, \sigma(D)) \geq \beta$, and **WRAcc**$(C, D) \geq 0$.

3   Related Work

In [3], we proposed an algorithm, named LOCOM, that computes the complete collection of rank-correlated subgroups. To enumerate a rank correlated subgroup (C, D), it first generates the part C, using an upper bound on the τ measure to discard unpromising patterns, and then find the subgroup D for which the correlation is very high. The first step is the most time consuming one and, in what follows, we propose to replace it with a sampling approach.

There are two families of local pattern sampling techniques. The first family is based on direct sampling approaches [4,5] that draw a sample without materializing auxiliary parts, but require a preprocessing step, the computation of weights on the whole set of observations, that makes it unusable when considering pairs of observations, like in our problem. The second one is based on Markov Chain Monte Carlo methods that performs a random walk over a transition graph representing the probability of reaching a pattern given the current one. This can be done with the guarantee that the distribution of the considered interestingness measure is proportional on the sample set to the one of the whole pattern set [6]. But the computational cost is very high, since the transition graph representing the probability of reaching a pattern given the current one, has to be materialized in both directions (generalization and specialization). Other approaches [7,8] relax this constraint. They are pragmatic approaches that, as we will see in the following, give very good results in practice.

4   Randomly Sampling Maximal Correlated Subgroups

We follow the approach proposed in [7], for randomly sampling graphs, also used in [8] for generating maximal itemsets. This method does not guarantee the uniformity of the sampling but is very efficient when the size of the set to sample is large as it is the case here. Hence, we want to sample the correlated part of the pattern, that is sets $C \in \mathcal{C}$. If we denote by k the size of $\mathcal{C}$, there are $\frac{3^k - 1}{2}$ possible ranked correlated patterns[1].

Algorithm 1. LoMax

Input: The data: $\mathcal{D} = \mathcal{O} \times \mathcal{C}$ and $\preceq$ the canonical order on $\mathcal{C} \times \{+, -\}$.
Output: A sampled rank correlated subgroup C.

1   **draw** a $\sim u([0, |\mathcal{C}|])$
2   $C \leftarrow \{a^+\}$
3   $Candidates \leftarrow \{x^s \in \mathcal{C} \times \{+, -\} \setminus C : \tau(x^s) \geq \beta\}$
4   **while** $|Candidates| \geq 0$ **do**
5      $sum \leftarrow 0$
6   **for** $x^s \in Candidates$ *ordered by* $\preceq$ **do**
7         $sum \leftarrow sum + p(C \cup x^s)$
8         $N[x^s] \leftarrow sum$
9   **draw** v $\sim u([0, 1])$
10   Find maximum x^s in N such that $v < N[x^s]$
11     $C \leftarrow \{C \cup x^s\}$
12     $Candidates \leftarrow \{x^s \in Candidates \setminus C : \tau(C \cup x^s) \geq \beta\}$

The sampling approach, see Algorithm 1, consists by starting with a positively signed attribute and gradually adding signed attributes with respect to a probability p that is proportional to the quality value of the obtained pattern: for a rank correlated pattern C, $p(C) \sim \frac{\tau(C)}{Z}$ with Z a normalization factor. The process stops when no signed attributes can be added without the value of τ being below the threshold. Once the rank-correlation pattern C obtained, its descriptions D, that maximize the **WRAcc** measure, are obtained using the enumeration algorithm of [3].

5   Experimental Results

For the experiments we used four real world datasets from the UCI database: (1) African heart disease that describes health and lifestyle ($\mathcal{C} = 6$, $\mathcal{R} = 1$) of 462 patients suffering of heart diseases; (2) Abalone that contains physical measurements of 4177 abalones ($\mathcal{C} = 6$, $\mathcal{R} = 2$); (3) Seismic-bumps that describes high energy seismic bumps in a coal mine (2584 observations, $\mathcal{C} = 14$, $\mathcal{R} = 5$); and (4) German credit a dataset where clients are described as good or bad credit risks (1000 observations, $\mathcal{C} = 15$, $\mathcal{R} = 6$). Note also that LOCOM and LoMax are implemented in C and the experiments run on a machine equipped with 8 Intel(R) Xeon(R) W-2125 CPU @ 4.00 GHz cores 126 GB main memory, running Debian GNU/Linux. To assess the quality of our sampling approach,

---

[1] For each correlated pattern of size s, there is 2^{s-1} signed patterns. As there are $\binom{k}{s}$ such patterns, we have $\frac{1}{2} \sum_{s \geq 0} \binom{k}{s} 2^k - 1 = \frac{3^n - 1}{2}$.

we compare LoMAX with LOCOM, the complete approach. We seek to answer the following questions: How does the random sample approximate the complete set of rank correlated subgroups? Are higher quality patterns more likely to be drawn? Do the extracted patterns cover all the data well? Is the computation of sample pattern efficient compared to the complete computation?

Quantitative Study. Figure 1 presents the proportion of sampled patterns with respect to the τ measure for different sample sizes (depending on the total number of rank-correlation patterns in the dataset). We can observe that LoMAX approximates well the global distribution of patterns along the τ measure. We can also see that small samples contain patterns of high τ value and therefore small samples may be sufficient to obtain high quality patterns. Notice that, we can observed two mixed distributions in Abalone dataset. This is due to some attributes that contains much more tiles than others which leads to patterns with lower correlations.

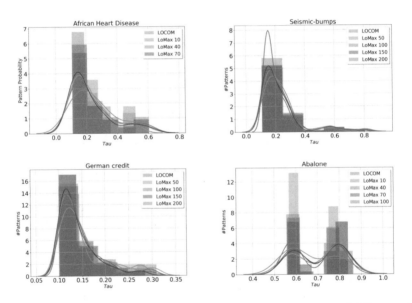

Fig. 1. Distribution of patterns (complete collection and samples of different sizes) with respect to their τ value: African Heart Disease (top left), Seismic-bumps (top right), German credit (bottom left) and Abalone (bottom right).

To study the effectiveness of the sampling method to find patterns with high τ values, we have randomly sampled a much larger number of patterns than there are in the dataset (the number computed by LOCOM). Hence, the same pattern may be drawn several times. Figure 2 reports, for each unique pattern, the number of times it has been drawn and we consider this number with respect to the pattern τ value. We can observe that the higher the τ value, the more the patterns are drawn. This shows that LoMAX favors patterns with high τ values, which is exactly what we are looking for.

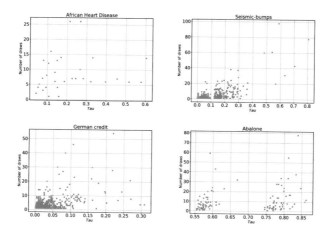

Fig. 2. Scatter plot of the number of identical patterns drawn with respect to τ. The sample size is equal to 10 times the number of patterns obtained with LOCOM.

In Fig. 3, we evaluate whether the extracted patterns cover the entire dataset or not. For this purpose, we compute Jaccard's measure between pairs of observations described by the patterns of the complete approach and those described by patterns of the random sample. More formally, let $\mathcal{M}_1$ be the set of maximal rank-correlated patterns extracted by LoMAX and $\mathcal{M}_2$ the one extracted by LOCOM. Let $N_i = \bigcup_{C \in \mathcal{M}_i} \eta(C, \mathcal{O})$. The Jaccard measure is thus $Jaccard(\mathcal{M}_1, \mathcal{M}_2) = \frac{N_1 \cap N_2}{N_1 \cup N_2}$: the higher the number of covered pairs, the higher the Jaccard value. Figure 3 (left) shows that a random sample of size about one third of the complete set of patterns is sufficient to cover a very large proportion of the pairs covered by LOCOM. Thus LoMAX generates very diverse patterns. Figure 3 (center) shows that the running time of LoMAX is lower than the one of LOCOM, and thus the computation of a sample of patterns is faster while offering a sample of very good quality patterns. Moreover, as each random draw is independent of the others, LoMAX can be run in parallel on several machines. We also study the **WRAcc** quality of the subgroups

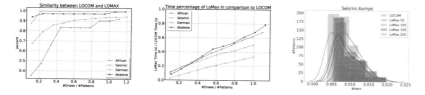

Fig. 3. (left) Jaccard value between the pairs of observations described by the patterns of the complete approach and those described by patterns of the random sample. (center) Running time of LoMAX divided by the one of LOCOM with respect to the number of draws divided by the number of patterns returned by LOCOM. (right) **WRAcc** value distribution for different sample sizes on Seismic-bumps dataset.

provided by LoMAX. Figure 3 (right) reports the distribution of the measure for seismic-bump dataset. It shows that the samples approximate well the complete collection of correlated subgroups.

Qualitative Study. LoMAX makes it possible to discover maximal correlated subgroups, especially to simultaneously highlight a data subspace and a rank-correlation pattern whose correlation is greater than that observed in the whole dataset. Due to some space limitation, we only report such patterns for Abalone and German credit datasets. For Abalone dataset, the top one drawn pattern with respect to the τ measure and the **WRAcc** is: $\langle \{Wweight^+, Shweight^+, length^+, diameter^+, height^+\}, \{sex = infant\}\rangle$ $(supp(D) = 0.321, \mathcal{T}(C, O) = 0.78, \mathcal{T}(C, \sigma(D)) = 0.8$, and **WRAcc**$(C, D) = 0.0025)$. This pattern means that, for *infant* abalones, the correlation between the whole weight, the shucked weight, the length of the longest shell, the diameter, and the height is stronger than on the whole dataset. Similarly, for German credit, we obtain $\langle \{Age^+, Duration^+, PrevCred^+\}, \{Housing = own\}\rangle$ $(supp(D) = 0.71, \tau(C, O) = 0.15, \mathcal{T}(C, \sigma(D)) = 0.18$, and **WRAcc**$(C, D) = 0.012)$. This pattern shows that the correlation between the attributes – Age, Duration and the number of existing credits – is weak but stronger for the people who are owner of their home.

6   Conclusion

In this paper, we study the problem of rank-correlated subgroup discovery that makes it possible to uncover subset of objects – identified by conditions on numerical and/or nominal attributes – for which the rank correlation among the attributes of an identified subset is exceptionally high. We propose a sampling algorithm to compute such pattern domain and study its performances. On four benchmark datasets we observe that the sampling approach reduces the computation cost, while identifying most important patterns. It makes also possible to process large databases by distributed computing on multiple machines. We think that the next direction could be to use the sampling for both the correlated pattern and the subgroup (description) to make the algorithm even faster.

Acknowledgements. This work was supported by the Labex IMU Université de Lyon (project RESALI) and a CNRS-INRIA CONFAP project.

References

1. Fayyad, U., Piatetsky-Shapiro, G., Smyth, P.: From data mining to knowledge discovery in databases. AI Mag. **17**(3), 37 (1996)
2. Lavrač, N., Flach, P., Zupan, B.: Rule evaluation measures: a unifying view. In: Inductive Logic Programming, pp. 174–185 (1999)
3. Hammal, M., Robardet, C., Plantevit, M.: Rank correlated subgroup discovery. J. Intell. Inf. Syst. (2019, to appear)

4. Boley, M., Lucchese, C., Paurat, D., Gärtner, T.: Direct local pattern sampling by efficient two-step random procedures. In: ACM SIGKDD, pp. 582–590 (2011)
5. Boley, M., Moens, S., Gärtner, T.: Linear space direct pattern sampling using coupling from the past. In: ACM SIGKDD, pp. 69–77 (2012)
6. Boley, M., Gärtner, T., Grosskreutz, H.: Formal concept sampling for counting and threshold-free local pattern mining. In: SDM, pp. 177–188 (2010)
7. Chaoji, V., Hasan, M.A., Salem, S., Besson, J., Zaki, M.J.: ORIGAMI. Stat. Anal. Data Min. **1**(2), 67–84 (2008)
8. Moens, S., Goethals, B.: Randomly sampling maximal itemsets. In: Workshop on Interactive Data Exploration and Analytics, IDEA 2013, pp. 79–86 (2013)

Multiple Sources Influence Maximization in Complex Networks with Genetic Algorithm

King Chun Wong[✉] and Kwok Yip Szeto[✉]

Department of Physics, The Hong Kong University of Science and Technology,
Clear Water Bay, Kowloon, Hong Kong
kcwongaz@connect.ust.hk, phszeto@ust.hk

Abstract. Information spreading is one of the most important classes of dynamical process in complex networks, as it has relevance in many applications in real-world spreading phenomena, such as the spreading of virus in epidemics, advertising through the diffusion of social opinion, and the cascade failure of power networks and financial systems. For the prevention and the control of epidemic, or for the advertisement in online marketing, it is important to search for the set of source nodes to serve as super carriers that can spread information most effectively over a given period of time. We first use a small Watts-Strogatz network to investigate the important features of the super carriers through exhaustive search. We then design a mutation-only genetic algorithm to search for these super carriers and compare the efficiency of genetic algorithm as well as the quality of the set of nodes in terms of a measure of influence in information spreading with exhaustive search. Finally, we extend this search method to a larger artificial network as well as a real network to provide a set of candidates super carriers.

Keywords: Information spreading · Genetic algorithm ·
Complex network · Combinatorial optimization

1 Introduction

Information spreading is an important class of dynamical process in networks that describes the spreading of influence from a set of information sources to the rest of the network and find uses in many different area, including marketing [1], power engineering [7], and many others. The central concerns in information spreading is often the identification of source nodes that maximize the diffusion of influence, so that either maximal influence can be produced or weak point to influence can be spotted for protection. This is known as the "influence maximization" problem. The simplest case of single source spreading had already been studied extensively in the literature [2]. As expected, the effectiveness of an individual source node in spreading information in the network is linked to the centrality of that node in the network and especially strongly correlated

© Springer Nature Switzerland AG 2020
F. Herrera et al. (Eds.): DCAI 2019, AISC 1003, pp. 226–234, 2020.
https://doi.org/10.1007/978-3-030-23887-2_26

with the node degree [6]. However, if one considers multiple sources spreading, because of the possible overlapping of the neighborhood of the source nodes, we expect that the best selection of source nodes will not simply be the set of nodes with the highest degree in the network, hence making the generalization of single source to multiple sources influence maximization non-trivial.

A large volume of literature on multiple-source information spreading focused on the problem of source detection, in which the goal is to identify the source nodes given an observation of a spreading process; some recent works include the K-center method [3] and the Gradient Maximum Likelihood Algorithm (GMLA) [8]. The influence maximization aspect of multiple-source spreading, on the other hand, is relatively less explored. Recently, Liu et al. has introduced a general-ized closeness centrality to extend the conventional closeness centrality to find optimal source sets in networks [5], but their construction is computationally expensive as it requires the computation of all distances in the network to the source nodes. In this paper, we propose a general method for identifying good candidate source set based on its topological features and a special genetic algo-rithm to search them effectively based on these features. We propose two key parameters computable from the local topology of the source nodes. From our observation that these two parameters are sufficient to capture most of the many-body effect in the spreading process, we design a special genetic algorithm to search for an optimal set of sources in large-scale real world networks using these two parameters. We demonstrate the efficiency of our method by comparing our result numerically with exhaustive search and extend our method to larger arti-ficial networks as well as some real networks.

2 Independent Cascade (IC) Model

2.1 Definition of Model

In this paper, we adopt the independent cascade (IC) model as our primary model for information spreading [9]. In the model, each node in the network can be in one of two states, either active or silent. The dynamics of the cascade process is defined by two assumptions. (1) Each active node i has only one single chance to activate (spread influence to) each of its silent neighbor j throughout the whole cascade process. (2) Each active node i activates its silent neighbor j by a diffusion probability Q_{ij} independent of all other diffusion probability Q_{kj} on node j by any other node k. Let's denote $A(t)$ to be the set of nodes that become activated at time step t. Under the above two assumptions, we run the cascade process as follows. Initially at time $t = 0$, all N nodes in the network is fixed to be in the silent state, except for a small number N_s of source nodes in the source set $A(t = 0)$. At each time step $t > 0$, each node i activated in the last time step ($i \in A(t-1)$) activates each of its silent neighbor by the diffusion probability Q_{ij}. We iterate this cascade process until either there is no more node to spread influence from, i.e. $A(t) = \emptyset$, or we reached a predefined duration length T. In this paper, we will focus on the case of uniform diffusion probability, $Q_{ij} = Q$. Having defined our model of cascade process, we can now rephrase our problem as to find the optimal set of source nodes $A(t = 0)$ such that the total number of nodes being activated at the end of the cascade process,

$$F(T; A(0), N, Q) = \left| \bigcup_{t=0}^{T} A(t) \right|, \tag{1}$$

is maximized. Because of the probabilistic nature of the cascade process, F needed to be averaged over some appropriate number of realizations.

2.2 Exact Solution and Observations

As mentioned in Sect. 1, because of the overlapping effect among source coverage, the degree centrality is no longer the only controlling parameter in the multiple source case. Physically, we may imagine each source node has a certain "radius of influence" proportional to its centrality measures, then the optimal set of source nodes will be the one with each of its member having the largest disk of influence while having these disks overlap the least. The above argument therefore motivates us to propose two parameters to be studied. The first parameter is called the group degree of the source set $A(t = 0)$ defined by

$$d = \sum_{j \notin A(0)} \left(\min \left\{ 1, \sum_{i \in A(0)} \mathcal{A}_{ij} \right\} \right), \tag{2}$$

where $\mathcal{A}$ is the adjacency matrix of the network. The group degree d measures the number of distinct neighboring nodes connected to the source set, which may be viewed as an estimate of the union of the disks of influence due to the sources. The second parameter is the group separation of the source set $A(t = 0)$,

$$l = \frac{1}{2} \sum_{i \in A(0)} \sum_{\substack{i \neq j, \\ j \in A(0)}} l(i, j), \tag{3}$$

where $l(i, j)$ is the shortest path length between nodes i and j. This group separation l is the sum of all pair-wise distances among the source nodes. We propose to use l as an estimate of the intersection of the disks of influence of individual source nodes, for it provides a measure of the mutual separations among the disks of influence. We conjecture that d and l are two good indicators in measuring the union of the disks of influence. We now proceed to test this conjecture by performing exhaustive search on small networks that is numerically treatable by exhaustive methods.

We first focus on the IC model defined on Watts-Strogatz (WS) networks for its importance in the study of social networks [12]. The WS network is defined by three parameters, its size N, average degree K, and rewiring probability P. We first consider the case of an IC process of $Q = 0.3, T = 10, N_s = 3$ on a WS network of $N = 50$, $K = 6$, $P = 0.3$, and compute the F-value of all $\binom{N}{N_s}$ possible source sets for 50 realizations of the WS network. We record the simulated F-value of each candidate source set together with their d and l values. We summarize the results in Fig. 1, which shows the density plot of the F values on the d and l plane. The grayscale of each point represents the average F-value of all source sets having that combination of d and l. We see a color gradient toward the upper right of the plot, which indicates that the best source set tends to have both high group degree and high group separation, in accord with our argument for introducing d and l.

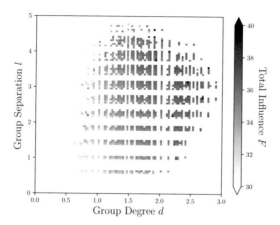

Fig. 1. Average influence F (represented by darkness) of source sets having different group degree d and group separation l. We see a gradient toward the upper right (higher d and l). The results produced from an IC process of $Q = 0.3, T = 10, N_s = 3$ on a WS network of $N = 50, K = 6, P = 0.3$.

Our insights so far are demonstrated only on small networks via exhaustive search, but exhaustive search is computationally impossible on much larger real networks, because of the $O(N\binom{N}{N_s})$ complexity in exhaustive simulation of all possible source sets. This therefore motivates our application of genetic algorithm to this problem. First of all, genetic algorithms are well known as a very effective optimization method in applications with large search space. Secondly, genetic algorithm provides us an "evolutionary history" of the source set, allowing us to seek correlation between our proposed measures and improvement of solution.

3   Genetic Algorithm

In this paper, we will adopt a specific class of genetic algorithm called mutation-only genetic algorithm (MOGA), in which the evolutionary actions are centered around a mutation operation [10]. (Do not confuse this with Multi-Objective Genetic Algorithm, which is also abbreviated as MOGA). Previous studies have shown that in many applications, mutation-only genetic algorithm can perform better than the standard simple genetic algorithms (SGA) [4, 10, 11], and this is especially true in network-optimization applications where simple crossover operators are more complex to operate. In this paper we use MOGA as an efficient method, by virtue of its success in other applications, to investigate cascade processes on very large real-world networks, and here we provide only a brief operational outline of MOGA without regard to question of its exact performance. For more details, please refer to the reference [4].

3.1   Operations of MOGA

Chromosome. Our candidate solution is a set of node indices indicating where to place the source nodes. We represent this set of indices by a chromosome, consisting of genes representing each individual node in the set. The population of our C candidate solutions is divided into three groups: (1) the current best group, (2) the current best individuals nodes, and (3) the randomly generated. These three groups respectively containing C_1, C_2, and C_3 chromosomes. Initially, we generate these chromosomes randomly.

Evaluation of Fitness. We evaluate the fitness of each chromosome by using the set of source nodes as sources in the IC model and compute the total number of activated node F due to the source (with appropriate averaging when $Q < 1$ in the IC model). We also keep record of the individual contribution by each source node to F, by breaking down the total information spread as $F = \sum_i f_i$, where f_i is the portion of the total spread due to the spread initiated by the i-th source node. We rank the nodes in terms of their individual fitness values f_i and store it to be used in the reproduction step.

Reproduction. We update the three groups of chromosomes separately. For the first group, we copy the C_1 chromosome having the highest fitness F. For the second group, we fill each chromosome with genes having the highest f_i values in the previous generation, in the hope that the individually best may also be the collectively best. This second group serves to fully exploit the information we gained in each generation. For each of the remaining C_3 chromosomes, we first fill half its genes by a random selection from the better half of all genes in the previous population. The remaining half are randomly generated. This third group hence provide the necessary exploration in the solution space.

4   Results

4.1   Preliminary Tests

We first use exhaustive search on small networks to obtain a benchmark for measuring the performance and quality of our results from genetic algorithm. To avoid the uncertainty in the fitness F intrinsic in the non-deterministic case ($Q < 1$) of information spreading, we first consider the deterministic case

Table 1. MOGA optimization results for five representative realizations of three test cases. In all three test cases, we used $K = 6$, $P = 0.3$, and $T = 3$. Exhaustive method results are listed on the left columns while MOGA results are listed on the right columns with subscript G. $G_{X\%}$ is the number of generations taken by the MOGA to reach at least $X\%$ of the exhaustive search solution fitness.

Test case	F	d	l	F_G	$G_{90\%}$	$G_{100\%}$	d_G	l_G
	226.0	17	3	226.0	40	769	17	3
$N = 250$	214.0	22	3	214.0	22	8190	22	3
$N_s = 2$	218.0	15	4	218.0	12	515	15	4
$Q = 1.0$	224.0	15	4	224.9	25	16991	15	4
	221.0	18	4	221.0	82	363	18	4
	44.433	21	8	44.733	2	2445	19	9
$N = 50$	44.433	18	7	44.9	4	197	22	8
$N_s = 3$	44.867	20	9	45.1	3	2299	20	7
$Q = 0.5$	44.533	20	7	45.467	3	94	23	8
	44.833	22	9	45.033	2	2657	22	9
	14.72	17	12	14.72	2	244	18	12
$N = 25$	14.9	19	12	15.04	5	1139	18	12
$N_s = 4$	14.92	21	16	14.98	6	3082	21	15
$Q = 0.2$	14.76	17	13	14.86	1	1679	18	14
	15.18	19	13	15.2	2	8865	19	14

of $Q = 1$. For the non-deterministic spreading, we repeat the experiment on two more test cases with different network and source sizes. We summarize the detailed computational results in Table 1 for five representative trials for each of the three test cases. We list the fitness value F, group degree d, and group separation l for both the exhaustive search and MOGA solutions. For the MOGA solution, we also record the number of generation G our MOGA needed to reach at least 90% and 100% of the exhaustive search solution fitness (denoted $G_{90\%}$ and $G_{100\%}$ respectively). We see that although our algorithm would not necessarily find exactly the same set of source nodes as in exhaustive search when $Q < 1$ because of the stochastic nature of the IC process, it can still find an alternative set that has a F-value at least as good. We should emphasize that for the small networks, our MOGA often can find solution having at least 90% of the global optimum in the first 100 generations already, as shown in Table 1. This means that for real network applications with very large N, the forbidden time complexity of exhaustive search can be avoided by our MOGA with relatively inexpensive computational resource. In the context of topological features, we note that the solutions found by MOGA have very similar group features (d and l) as the solution from exhaustive search, again echoing our finding in Sect.2.

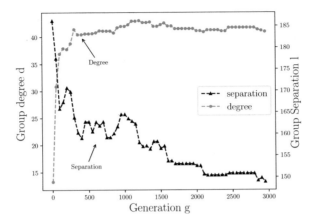

Fig. 2. Evolution of the MOGA solution for the power grid network, averaged over 20 trials. The upper and lower curves recorded respectively how the group degree d and group separation l evolved with the evolving MOGA solution.

4.2   Application to Real Data

To test the validity of our conclusions from last section in large-scale networks, typically having thousands, or even millions of nodes, we set up an IC model on a real-world networks and apply MOGA to find the optimal source set. The network we used is the Western States Power Grid [12], with a large number of nodes ($N = 4941$). We choose for this numerical test $N_s = 5$, $Q = 0.3$, and $T = 10$ and average the result over 20 trial runs. Figure 2 shows the results. As the MOGA solution evolves to higher fitness, the solution source set tends towards having higher group degree, agreeing with our previous observations. The group separation, however, shows a decreasing trend instead, which seems to contradict our earlier arguments in Sect. 2. This opposite trend of the two curves in Fig. 2 suggests that for the power grid network, the group separation l is relatively less important than the group degree d, so that it is favorable to give up some separation for more degree centrality. For real networks lacking the high degree of clustering as in the WS network model, it is possible that there are many "isolated" nodes very far separated from most other nodes in the network so it is more favorable to choose node within the major cluster in the network as influence sources. Following this argument, the decreasing trend of the group separation may not pose a real contradiction on the importance of l but instead is a consequence to our need to simultaneously maximize d and l.

5   Conclusion

In this paper, we studied the multiple source influence maximization problem in complex networks. We observed that in small network models, the quality of the source set is largely determined by two local topological features, namely the group degree and group separation. We designed a special version of mutation-only genetic algorithm to replace exhaustive methods which cannot be practical for large networks. We verified the efficiency of our MOGA in finding the optimal source sites in small networks of size $N \sim 100$ and found that MOGA can achieve over 90% of the global maximum efficiently using a much smaller computational resource than exhaustive search. By applying our MOGA to a real network of size more than a few thousands nodes, we observed that group degree is more important than group separation in information cascade for our sample network, possibly due to its low connectivity. Nevertheless, we suggest that degree centrality should still be a major factor in information spreading for denser networks, e.g. networks with high average clustering coefficient. For future research, it will be interesting to extend our analysis and algorithm to other information spread model beyond the simple IC model, as to capture more complicated communication dynamics.

Acknowledgments. K. C. Wong acknowledges the support of UROP funding from HKUST.

References

1. Domingos, P.: Mining social networks for viral marketing. J. Retail. Consum. Serv. **20**, 80–82 (2005)
2. Jalili, M., Perc, M.: Information cascades in complex networks. J. Complex Netw. **5**, 665–693 (2017)
3. Jiang, J., Wen, S., Yu, S., Xiang, Y., Zhou, W.: K-center: an approach on the multi-source identification of information diffusion. IEEE Trans. Inf. Forensics Secur. **10**(12), 2616–2626 (2015)
4. Law, N.L., Szeto, K.Y.: Adaptive genetic algorithm with mutation and crossover matrices. In: Proceedings of the 20th International Joint Conference on Artifical Intelligence (IJCAI 2007), Hyderabad, India, pp. 2330–2333, January 2007. http://dl.acm.org/citation.cfm?id=1625275.1625651
5. Liu, H.L., Ma, C., Xiang, B.B., Tang, M., Zhang, H.F.: Identifying multiple influential spreaders based on generalized closeness centrality. Phys. A: Stat. Mech. Appl. **492**, 2237–2248 (2018)
6. Lu, L., Chen, D., Ren, X.L., Zhang, Q.M., Zhang, Y.C., Zhou, T.: Vital nodes identification in complex networks. Phys. Rep. **650**, 1–63 (2016)
7. Motter, A.E., Lai, Y.C.: Cascade-based attacks on complex networks. Phys. Rev. E **66**(6), 065102 (2002)
8. Paluch, R., Lu, X., Suchecki, K., Szymanski, B., Holyst, J.: Fast and accurate detection of spread source in large complex networks. Sci. Rep. **8**, 2508 (2018)
9. Saito, K., Nakano, R., Kimura, M.: Prediction of information diffusion probabilities for independent cascade model. Knowl.-Based Intell. Inf. Eng. Syst. Lect. Notes Comput. Sci. **5179**, 67–75 (2008)
10. Szeto, K.Y., Zhang, J.: Adaptive genetic algorithm and quasi-parallel genetic algorithm: application to knapsack problem. In: Large-Scale Scientific Computing, pp. 189–196. Springer, Berlin (2006)

11. Wang, G., Wu, D., Chen, W., Szeto, K.Y.: Importance of information exchange in quasi-parallel genetic algorithms. In: Proceedings of the 13th Annual Conference Companion on Genetic and Evolutionary Computation, GECCO 2011, pp. 127–128. ACM, New York (2011)
12. Watts, D.J., Strogatz, S.H.: Collective dynamics of 'small-world' networks. Nature **393**, 440 (1998)

Detecting Topics in Documents by Clustering Word Vectors

Guilherme Raiol de Miranda[1,2(✉)], Rodrigo Pasti[2],
and Leandro Nunes de Castro[1,2]

[1] Natural Computing Laboratory, Mackenzie Presbyterian University,
São Paulo, Brazil
lnunes@mackenzie.com.br
[2] AxonData Tecnologia Analítica, São Paulo, Brazil
{guilherme.miranda,rodrigo.pasti}@axondata.com.br

Abstract. The automatic detection of topics in a set of documents is one of the most challenging and useful tasks in Natural Language Processing. Word2Vec has proven to be an effective tool for the distributed representation of words (word embeddings) usually applied to find their linguistic context. This paper proposes the use of a Self-Organizing Map (SOM) to cluster the word vectors generated by Word2Vec so as to find topics in the texts. After running SOM, a k-means algorithm is applied to separate the SOM output grid neurons into k clusters, such that the words mapped into each centroid represent the topics of that cluster. Our approach was tested on a benchmark text dataset with 19,997 texts and 20 groups. The results showed that the method is capable of finding the expected groups, sometimes merging some of them that deal with similar topics.

Keywords: Topic detection · Word2Vec · Self-Organizing Maps · K-Means · Word embeddings

1 Introduction

The generation of new data has grown dramatically in recent years, reaching a mark of 2.5 exabytes per day [1]. Much of this data is unstructured, such as images, videos, and texts. Due to the high reach of the Internet in society, millions of texts are produced every day on social networks like Twitter and Facebook, which makes it increasingly difficult for humans to interpret what people are talking about. This paper proposes the first step towards the creation of a tool that allows automatic word-based topic detection without the need for prior knowledge of the texts and number of topics, and that can be scalable to the large amount of content produced in the present day.

In recent years, *Word2Vec* emerged as one of most promising tools in Natural Language Processing. The algorithm represents the words in a n-dimensional vector space, while maintaining the semantic and syntactic relations of the words

© Springer Nature Switzerland AG 2020
F. Herrera et al. (Eds.): DCAI 2019, AISC 1003, pp. 235–243, 2020.
https://doi.org/10.1007/978-3-030-23887-2_27

from the input texts [2]. From the vectors one can apply techniques that capture the similarities by means of vector operations [3] or cluster the word vectors.

To perform data visualization with dimensionality reduction a Self-Organizing Map (SOM), also known as Kohonen Map, was used. A SOM is a self-organizing neural network with a simple architecture (consisting of one- or two-dimensional output grid), where each unit competes for an n-dimensional input vector and generates a lower-dimensional map that represents the topology of the input data [4].

The research hypothesis of this paper is the following: the combination of SOM with k-means and Word2Vec generates a useful tool for the extraction of topics from documents with various subjects.

The work is structured as follows. Section 2 presents the background to the tools used; Sect. 3 presents the related work; Sect. 4 contains the proposed methodology for solving the problem; Sect. 5 shows the results and analysis; and Sect. 6 provides the conclusions and future works.

2   Background

The world lives a revolution in the way data is generated and consumed, being treated as the Fourth Industrial Revolution, commonly called Industry 4.0. The exponential growth in data generation and the increased processing power of computers have created many markets and boosted other already traditional ones in the academy, such as Artificial Intelligence (AI), Machine Learning (ML), and Natural Language Processing (NLP). Algorithms in these areas are needed mainly when there is a large dataset and/or when data is unstructured. The three main techniques explored in this research are: Word2Vec; Self-Organizing Maps; and K-Means.

The *Word2Vec* is one of the most promising techniques in NLP that consists of a model to represent words as vectors (word vectors) in a way that words used together tend to stay closer in the vector space. Word2Vec is a Neural Network trained using word sequences, centered on a specific word, in order to maximize the average logarithmic probability that each word of the sequence appears along with the central word, using a word context of k words [2] (named window). As output, n-dimensional vectors are generated (called *word vectors*) for each input word, and each vector will be close in the vector space to the representations of the words that tend to be used together.

Furthermore, vector representations carry various linguistic regularities and patterns, and these relations can be found with vector operations, such as linear translations [2]. A classic example is the use of word vectors for analogies between words: "**man** is for **woman** as king is for **x**?". Using simple vector operations, the answer could be obtained by the equation $x \approx king - man + woman$, where **x** would approximate the vector representing the word **queen** [3].

While the Self-Organizing Map is also a type of neural network, its architecture differs from Word2Vec by being based on similar processes that occur in the human brain (such as projections in sensory areas), consisting of a one- or two-layer neural network with unsupervised learning in which neurons compete to better adapt to the inputs [4]. At the end of the training process, neurons maintain a topological order similar to that of the input, or there will be a loss of information. Redundancy will be eliminated, facilitating the processing in later stages [5]. The output of the Map is a one- or two-dimensional matrix, which can be represented using a U-Matrix (unified distance matrix), where the topology of the input data is represented by color levels or contours.

Finally, *K-Means* is an unsupervised algorithm for clustering data. By defining a number of groups, K, the algorithm will initially position K random points, called centroids, and the groups will be formed by the objects (data) closest to the centroids. At each iteration it is calculated the error given by the distance between the group's objects and the centroid, updating the position of the centroid according to the error.

3   Related Work

Since the development of Word2Vec, there have been several works in the literature exploring its characteristics and applying it in various situations with different techniques. Two of these works combined the Word2Vec model with Self-Organizing Maps [6,7], but the generated word vectors were used to improve the text representation for document clustering. Also, there are studies that have used word vectors clustering as a step to cluster documents [8] and word vectors within each document [9].

One way to use SOM for contextual recognition has been proposed by [10]. Named Semantic Self-Organizing Maps, it divides the neurons based on their roles, in which each word would be presented to the neural network using its predecessors and immediate successors to determine context. The experiments performed were limited and only with artificial data.

This paper proposal is closer to the Latent Dirichlet Allocation (LDA) approach. The LDA algorithm is a probabilistic topic modelling method that performs the task of identifying topics that best describe a set of documents [11].

4   The Proposed Approach

The goal of this paper is to propose a method to find topics in text documents by clustering word vectors using a Kohonen Self-Organizing Map and k-means. The proposed approach is composed of the following five steps:

1. Data collection and selection;
2. Text Pre-Processing;
3. Generation of word vectors;
4. SOM generation; and
5. Clustering with k-means and topic detection.

4.1 Data Collection and Selection

The data selected to assess the proposed method was the *Twenty Newsgroup Dataset*, donated to the Machine Learning repository of the University of California, Irvine (UCI) [12]. The set contains twenty e-mail discussion groups, containing approximately 1,000 emails in each group, separated in the following subjects: Atheism, Computer Graphics, Windows, Windows-Misc, IBM hardware, Mac Hardware, Cars, Guns, Motorcycles, Baseball, Hockey, Cryptography, Electronics, Medicine, Space, Christianity, Weapons, Middle East, Politics and Religion.

4.2 Text Pre-processing

The following text pre-processing techniques were used: special characters removal; transformation of all words into lower case; tokenization; stopwords removal; and removal of common words, such as verbs, nouns and adverbs.

4.3 Word Vectors' Generation

To generate the word vectors we used the implementation of Word2Vec in the *Gensim* package for Python. The parameters used in the generation of word vectors were: 50 minimum word occurrences, 10 words in the window for analysis (10 previous and subsequent words for a given target), and word vectors with 50 dimensions.

4.4 SOM Generation

The *Sompy* package was used to implement the SOM algorithm. It allows the visualization of the features separated by dimensions, the visualization of the map contours and has a grouping function already implemented. It was used the standard implementation parameters: rectangular topology and Gaussian neighborhood function.

4.5 Data Clustering and Topic Detection

The *K-Means* implementation within Sompy was used. The neurons were grouped in the same number of input subjects. After grouping, for each word it was assigned to a best-matching neuron, and the word topic is the same as the neuron that represents it.

5 Performance Evaluation

To verify if the hypothesis question could be validated with the proposed hybrid method, it was made a preliminary experiment with three subjects: Atheism, Hockey and Windows, totaling 3,000 texts. If the first experiment presents a good performance with diverse subjects and is able to form groups of topics, then further experiments can be performed to find topics in all twenty subjects and 19,997 texts.

The preliminary experiment generated 938 word vectors that were used in a 10×10 SOM grid and grouped into 3 topics by k-Means. The experiment with the full dataset generated 5,175 word vectors as input to a 30×30 SOM grid, grouped in 20 topics with k-means. The rationale of these parameters was to use the dimensionality reduction capacity of SOM to reduce redundant data and verify if the groups formed by k-means would represent topics similar to the input subjects. Thus, the number of input subjects was used to define the number of centroids in k-means.

Figure 1a shows the U-Matrix generated by the SOM output grid for the preliminary experiment. By observing this picture, it is possible to note three different groups: a dark blue group with high level curves, a light blue group, and a reddish group. This can be confirmed by applying k-means to cluster the SOM output neurons, as presented in Fig. 1b.

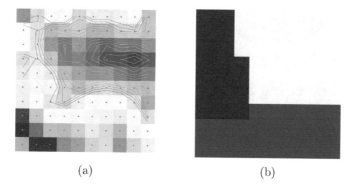

(a)   (b)

Fig. 1. SOM U-Matrix (a) and Decision Map (b) with three subjects

As there was a decrease in data redundancy in the SOM mapping task, each neuron represents several word vectors. The distribution of neurons, the number of words per topic and some sample words of each topic can be seen in Table 1.

Table 1. SOM Results with Three Topics

Topic	Neurons	Words	Example of words in each topic
0	39	280	memory, distribution, intel, system, internet
1	36	402	christian, morality, theory, atheists, creation
2	25	256	americans, Pittsburgh, saves, scorer, season

By analyzing the grouped words in Table 1, one can observe all entry topics: Atheism, Hockey and Windows. Topic 0 has several words associated with computers and emails, characterizing the topic **Windows**. Topic 1 has several words about religion, science, purpose, etc., which represent the topic **Atheism**. Finally, Topic 2 has several names of teams, places and words related to games, corresponding to the topic **Hockey**.

For the second experiment with the full data, the SOM's results are presented in Fig. 2a, and the topics found by the application of k-means to the SOM's output neurons are presented in Fig. 2b.

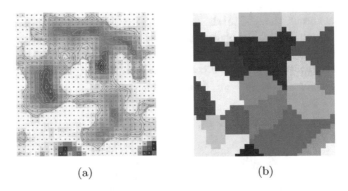

(a)   (b)

Fig. 2. SOM U-Matrix (a) and Decision Map (b) with twenty subjects

Each output topic is presented with its 5 most frequent words in Table 2. To verify the quality of the output topics found, the following method was proposed: take all words belonging to one output topic, count the number of occurrences on each entry topic and then divide by the absolute frequency of words on each output topic. Entry topics with more than a 10% of relevance are considered dominant for the output topic.

By analyzing the topics created, some important observations can be made:

- Some topics contain related subjects that result in similar labels. This happened with topics related to computing (Topics 3, 11 and 19), religion (Topics 6 and 16) and sports (Topic 7).
- Some topics have relatively few words but represent a large set of data (Topics 8 and 12). By observing the words, it is possible to note that they are related to sending and receiving e-mail, and since these words are usually in almost all e-mails' headers, they have a high frequency of occurrence. Following this logic, there were topics with large numbers of words related to the exchange of e-mails (Topics 1 and 10), but that did not have any label. This happens because, although they are words that represent the exchange of e-mails, they are words that appear in only a few sub-threads.

Table 2. SOM results with Twenty Topics Caption for Dominant Topics: Atheism [1], Autos [2] Baseball [3], Christianity [4], Computer Graphics [5], Cryptography [6], Electronics [7], For Sale [8] Guns [9], Hockey [10], IBM hardware [11], Mac Hardware [12], Medicine [13], Middle East [14], Motorcycles [15], Politics [16], Religion [17], Space [18], Windows [19], Windows-Misc [20]

Topic	Neurons	Words	Most frequent words	Dominant topics
0	35	164	time, year, years, free, order	-
1	68	375	space, school, york, division, buffalo	-
2	51	294	right, open, hand, side, model	2, 15
3	27	187	system, problem, hardware, drive, apple	20, 12, 11
4	62	341	people, good, anyone, something, much	2, 15, 9, 7, 1, 13, 17, 16, 3, 4
5	54	249	research, national, president, states, house	16
6	56	393	point, question, fact, true, reason	1, 17, 4
7	57	296	game, second, team, games, hockey	10, 3
8	11	68	cantaloupe, state, newsgroups, path, ohio	20, 5, 2, 15, 12, 19, 8, 10, 7, 13, 3, 11, 4
9	59	402	many, different, high, possible, human	13
10	78	423	subject, talk, david, john, bill	-
11	34	146	windows, mail, thanks, file, graphics	20, 5, 19
12	15	73	news, lines, date, message, organization	20, 5, 2, 18, 15, 12, 19, 8, 10, 7, 13, 6, 3, 11
13	33	252	religion, american, turkish, israel, jews	14
14	34	167	first, life, children, times, home	-
15	54	341	power, small, speed, black, white	7, 18
16	47	328	christian, jesus, word, bible, history	4, 17
17	42	184	information, number, part, please, group	6
18	31	200	government, case, person, rights, control	9, 16
19	52	292	data, line, chip, window, type	6, 19

6   Conclusions and Future Works

Finding topics in text has been dealt with using different approaches, from simply counting the frequency of words to using sophisticated machine learning algorithms. In this paper we proposed a hybrid, cascade system, in which a first method to build word embeddings is used to generate distributed word representations, and, then, a self-organizing network maps the word vectors into regiones of similar words in the SOM output grid and then a k-means algorithm segments the SOM neurons in the search for clusters of data (topics).

To test the proposed approach we used a real-world dataset with 19,997 different texts distributed over 20 clusters. By assigning the number of topics equals to that of the input subjects, the results of the grouping task generated topics of words that resembled the contexts of each subject. This suggests that the proposed approach was able to find the topics in the documents as labelled by a human subject.

Future work includes the study of other grouping techniques that do not require previous knowledge of the number of groups and the comparison with other redundancy elimination algorithms for data preparation, such as an Artificial Immune Network [13]. Also, further research will assess the computational scalability of the method when using High Performance Computing (HPC) platforms. New experiments will be performed with large datasets, as those available in social media.

Acknowledgments. The authors thank CAPES, CNPq, Fapesp, Mackpesquisa and Intel for the financial support.

References

1. McAfee, A., Brynjolfsson, E., Davenport, T.H., Patil, D., Barton, D.: Big data: the management revolution. Harvard Bus. Rev. **90**(10), 60–68 (2012)
2. Mikolov, T., Sutskever, I., Chen, K., Corrado, G.S., Dean, J.: Distributed representations of words and phrases and their compositionality. In: Advances in Neural Information Processing Systems, pp. 3111–3119 (2013)
3. Levy, O., Goldberg, Y.: Linguistic regularities in sparse and explicit word representations. In: Proceedings of the Eighteenth Conference on Computational Natural Language Learning, pp. 171–180 (2014)
4. Kohonen, T.: Self-organized formation of topologically correct feature maps. Biol. Cybern. **43**(1), 59–69 (1982)
5. de Castro, L.N., et al.: Análise e síntese de estratégias de aprendizado para redes neurais artificiais (1998)
6. Yoshioka, K., Dozono, H.: The classification of the documents based on word2vec and 2-layer self organizing maps. Int. J. Mach. Learn. Comput. **8**(3) (2018)
7. Subramanian, S., Vora, D.: Unsupervised text classification and search using word embeddings on a self-organizing map. Int. J. Comput. Appl. **156**(11) (2016)
8. Shi, M., Liu, J., Zhou, D., Tang, M., Cao, B.: WE-LDA: a word embeddings augmented LDA model for web services clustering. In: 2017 IEEE International Conference on Web Services (ICWS), pp. 9–16. IEEE (2017)

9. Dai, X., Bikdash, M., Meyer, B.: From social media to public health surveillance: word embedding based clustering method for Twitter classification. In: Southeast-Con 2017, pp. 1–7. IEEE (2017)

10. Ritter, H., Kohonen, T.: Self-organizing semantic maps. Biol. Cybern. **61**(4), 241–254 (1989)

11. Blei, D.M., Ng, A.Y., Jordan, M.I.: Latent Dirichlet allocation. J. Mach. Learn. Res. **3**(Jan), 993–1022 (2003)

12. Mitchell, T.: UCI machine learning repository (1999)

13. de Castro, L.N., Von Zuben, F.J.: aiNet: an artificial immune network for data analysis. In: Data Mining: A Heuristic Approach, pp. 231–260. IGI Global (2002)

A Forward-Backward Labeled Multi-Bernoulli Smoother

Rang Liu, Hongqi Fan, and Huaitie Xiao[✉]

National Key Laboratory of Science and Technology on ATR,
College of Electronic Science, National University of Defense Technology,
Changsha, Hunan 410073, China
liurang13@163.com, fanghongqi@nudt.edu.cn,
htxiao@126.com

Abstract. A forward-backward labeled multi-Bernoulli (LMB) smoother is proposed for multi-target tracking which includes a forward LMB filtering part and a backward LMB smoothing part. The forward LMB filtering is the standard LMB filter, and the backward LMB smoothing recursion is closed. The derived backward smoothed density is a LMB distribution under LMB prior.

Keywords: Random finite set · Smoother · Labeled multi-Bernoulli

1 Introduction

There are three types of estimation techniques for target tracking [1], i.e., prediction, filtering and smoothing where the smoothing usually estimates the previous target state with all measurements till the current time. Because more measurements are used for estimation, the smoothing has better performance than filtering. For multi-target tracking application, the tracking is troubled by clutter, missed detection and data association uncertainty [2]. The method based on random finite sets (RFS) can better integrate these uncertainties into the filter framework, and has become a hot topic in recent years. The goal of this paper concerns on the smoothing in the context of RFS filtering.

The forward-backward probability hypothesis density (PHD) smoother is proposed in [3–5]. The PHD smoother can improve the accuracy of the position estimation but cannot necessarily lead to a better cardinality estimation. The Bernoulli smoother is proposed in [6]. The smoother performs better than the Bernoulli filter, but the Bernoulli smoother adapts to at most one target. A forward-backward Cardinality Balanced multi-Bernoulli (CBMeMBer) smoother is proposed in [7], its SMC implementation of CBMeMBer smoother has a high complexity and may cause that the death target is dropped before a target dies. The generalized labeled multi-Bernoulli (GLMB) smoother is proposed in [8] which is the first closed form solution to the smoothing with labeled RFS, but the smoother have a complicated data association so it has not been implemented in practice.

Inspired by the effort in [6, 8], in this paper we derive an exact forward-backward labeled multi-Bernoulli (LMB) smoother for multi-target tracking. The forward LMB

F. Herrera et al. (Eds.): DCAI 2019, AISC 1003, pp. 244–252, 2020.
https://doi.org/10.1007/978-3-030-23887-2_28

filtering part is the standard LMB filter. Our work mainly concerns on the backward smoothing part. We prove that the backward smoothing recursion is closed under the LMB prior for the standard multiple target models and the backward smoothed density of each target is similar to the Bernoulli backward smoothed density in the form of [6].

The rest of the paper is organized as follows. We review the basic definitions of labeled RFS and introduce the multi-target Bayes forward-backward smoother in Sect. 2. In Sect. 3, we derive a forward-backward LMB smoother. The conclusion is in Sect. 4.

2 Background

2.1 Labeled RFS

State space is represented by blackboard bold letter, for example, the label space is represented as $\mathbb{L}$, and the unlabeled target state space is represented as $\mathbb{X}$. Single target state and multi-target state are represented by lowercase letter and uppercase letter respectively. x and X are the unlabeled state representations. $\mathbf{x}$ and $\mathbf{X}$ are the labeled state representations. The labeled single target state has the form $\mathbf{x} = (x, \ell)$, and $x \in \mathbb{X}, \ell \in \mathbb{L}$. A labeled multi-target state has the form $\mathbf{X} = \{\mathbf{x}_1, \ldots, \mathbf{x}_i, \ldots, \mathbf{x}_{|\mathbf{X}|}\}$, where $|\mathbf{X}|$ represents multi-target cardinality (or the number of targets), $\mathbf{x}_i$ represents a labeled single target state, and $\mathbf{X} \subset \mathbb{X} \times \mathbb{L}$. Define a projection $L : \mathbb{X} \times \mathbb{L} \to \mathbb{L}$ to make $L(\mathbf{x}) = \ell$ and $L(\mathbf{X}) \triangleq \{\ell | \mathbf{x} \in \mathbf{X}\}$. The inner product of $f(\mathbf{x})$ and $g(\mathbf{x})$ on $\mathbb{X} \times \mathbb{L}$ is defined as $\langle f, g \rangle = \int f(\mathbf{x}) g(\mathbf{x}) d\mathbf{x}$. The multi-target exponential form $h^{\mathbf{X}}$ is defined as $\prod_{\mathbf{x} \in \mathbf{X}} h(\mathbf{x})$ with $h^{\varnothing} = 1$. Then we define the inclusion function as:

$$\delta_{\mathbf{Y}}(\mathbf{X}) \triangleq \begin{cases} 1, & \mathbf{X} = \mathbf{Y} \\ 0, & otherwise \end{cases}, 1_{\mathbf{Y}}(\mathbf{X}) \triangleq \begin{cases} 1, & \mathbf{X} \subset \mathbf{Y} \\ 0, & otherwise \end{cases} \tag{1}$$

When $\mathbf{X}$ has different labels, $\Delta(\mathbf{X}) \triangleq \delta_{|\mathbf{X}|}(|L(X)|) = 1$. The LMB distribution is ([2])

$$\pi(\mathbf{X}) = \Delta(\mathbf{X}) \omega(L(\mathbf{X})) p^{\mathbf{X}}, \omega(I) = \prod_{\ell \in \mathbb{L}} (1 - r^{\ell}) \prod_{\ell \in I} \frac{1_{\mathbb{L}}(\ell) r^{\ell}}{1 - r^{\ell}} \tag{2}$$

Where r^{ℓ} represents the existence probability of the track ℓ, $p(\cdot, \ell)$ represents the probability density and $\omega(I)$ is the weight of the hypothesis $I = \{\ell_1, \ldots, \ell_{|I|}\}$. The above LMB distribution is the special case of GLMB distribution in [7].

The probability densities and the existence probabilities of different targets in LMB RFS are both uncorrelated. If $\mathbf{X}_1$ and $\mathbf{X}_2$ with $\mathbf{X}_1 \subset \mathbb{X} \times \mathbb{L}_1, \mathbf{X}_2 \subset \mathbb{X} \times \mathbb{L}_2$ and $\mathbb{L}_1 \cap \mathbb{L}_2 = \varnothing$ are both LMB RFS, $\mathbf{X} = \mathbf{X}_1 \cup \mathbf{X}_2$ is a LMB RFS and vice versa. Moreover, the probability densities of $\pi_1(\mathbf{X}_1)$, $\pi_2(\mathbf{X}_2)$ and $\pi(\mathbf{X})$ have the relations:

$$\pi(\mathbf{X}) = \pi_1(\mathbf{X}_1) \pi_2(\mathbf{X}_2) \tag{3}$$

2.2   Multi-target Bayes Forward-Backward Smoother and Motion Model

The multi-target Bayes forward-backward smoother includes two parts, i.e., the forward filtering and the backward smoothing. The prediction and update equations for the forward filtering can be obtained from [1–3]. The recursion equation of backward smoothed density $\pi_{k-1/t}(\mathbf{X})$ can be represented as [3–8]

$$\pi_{k-1/t}(\mathbf{X}) = \pi_{k-1/k-1}(\mathbf{X}) \int f_{k/k-1}(\mathbf{Y}|\mathbf{X}) \frac{\pi_{k/t}(\mathbf{Y})}{\pi_{k/k-1}(\mathbf{Y})} \delta \mathbf{Y} \tag{4}$$

Where $\pi_{k/t}(\mathbf{Y})$ is the backward smoothed density from moment t to moment k ($k \le t$) and $\pi_{k-1/t}(\mathbf{X})$ is the backward smoothed density from moment t to moment $k-1$. $\pi_{k-1/k-1}(\mathbf{X})$ is the multi-target posterior density at moment $k-1$ and $\pi_{k/k-1}(\mathbf{Y})$ is the predicted multi-target density at moment k. $f_{k/k-1}(\mathbf{Y}|\mathbf{X})$ is the Markov transition density at moment k and will be introduced as follow.

Let single target Markov transition density $f_{k/k-1}(x_+|(x,\ell))$ represents the probability that the target (x,ℓ) at moment $k-1$ transits to (x_+,ℓ) at moment k. $\mathbf{Y}^-$ denotes the surviving multi-target state from moment $k-1$ to moment k and is a LMB RFS with the parameterized form $\left\{ \left(p^\ell_{s,k/k-1}, f_{k/k-1}(x_+|(x,\ell)) \right) \right\}_{\ell \in \mathbf{X}}$ where $\mathbf{Y}^- \subseteq \mathbb{X} \times \mathbb{L}$ and $\mathbb{L} = \mathbb{L}_{1:k-1}$. When the newborn targets are not considered, multi-target Markov transition density $f_{s,k/k-1}$ can be written as

$$f_{s,k/k-1}(\mathbf{Y}^-|\mathbf{X}) = \Delta(\mathbf{Y}^-)\Delta(\mathbf{X}) \times$$
$$\left(1 - p_{s,k/k-1}(x,\ell)\right)^{\mathbf{X}} \prod_{\ell \in L(\mathbf{Y}^-)} \frac{1_{L(\mathbf{X})}(\ell) p_{s,k/k-1}(x,\ell) f_{k/k-1}(y|x,\ell)}{\left(1 - p_{s,k/k-1}(x,\ell)\right)} \tag{5}$$

It is assumed that the newborn targets can be represented by a LMB RFS denoted as $\mathbf{Y}^+$ where $\mathbf{Y}^+ \subseteq \mathbb{X} \times \mathbb{B}$ and $\mathbb{B} = \mathbb{L}_k$. The density $f_{B,k/k-1}$ of newborn targets can be represented as

$$f_{B,k/k-1}(\mathbf{Y}^+) = \Delta(\mathbf{Y}^+) \prod_{\ell \in \mathbb{B}} \left(1 - r^\ell_{B,k/k-1}\right) \prod_{\ell \in L(\mathbf{Y}^+)} \frac{1_{\mathbb{B}}(\ell) r^\ell_{B,k/k-1} p_{B,k/k-1}(y,\ell)}{\left(1 - r^\ell_{B,k/k-1}\right)} \tag{6}$$

The multi-target state at moment k can be represented as $\mathbf{Y} = \mathbf{Y}^+ \uplus \mathbf{Y}^-$ where $\mathbf{Y}^+$ and $\mathbf{Y}^-$ are disjoint and independent. From (3), the total multi-target Markov transition density $f_{k/k-1}$ can be represented as

$$f_{k/k-1}(\mathbf{Y}|\mathbf{X}) = f_{B,k/k-1}(\mathbf{Y}^+) f_{s,k/k-1}(\mathbf{Y}^-|\mathbf{X}) \tag{7}$$

3 Forward-Backward LMB Smoother

The multi-target Bayes smoother is usually computationally intractable. A Forward-Backward LMB smoother which tractably approximates the multi-target Bayes smoother is derived. The forward LMB filtering is the standard LMB filter, and see [2] for details. Next, we will prove that the backward smoothing is closed under LMB prior.

Proposition 1: For the forward-backward LMB smoother, when the forward filtering is running to the moment t, the multi-target posterior density $\pi_{k-1/k-1}(\mathbf{X})$ at moment $k-1$ and the predicted multi-target density $\pi_{k/k-1}(\mathbf{Y})$ at moment k are both approximated by LMB RFS ($k \leq t$). If the backward smoothed density from moment t to moment k is a LMB RFS and can be represented as $\pi_{k/t}(\mathbf{Y})$, we can obtain that the backward smoothed density $\pi_{k-1/t}(\mathbf{X})$ from moment t to moment $k-1$ is also a LMB RFS and can be written as

$$\pi_{k-1/t}(\mathbf{X}) = \left\{ \left(r^\ell_{k-1/t}, p_{k-1/t}(x, \ell) \right) \right\}_{\ell \in \mathbb{L}} \tag{8}$$

Where

$$r^\ell_{k-1/t} = 1 - \frac{\left(1 - r^\ell_{k-1/k-1}\right)\left(1 - r^\ell_{k/t}\right)}{\left(1 - r^\ell_{k/k-1}\right)} \tag{9}$$

$$p_{k-1/t}(x, \ell) = $$
$$\frac{p_{k-1/k-1}(x, \ell)\left(\alpha_{s,k/t}(x, \ell) + \beta_{s,k/t}(x, \ell)\int \frac{f_{k/k-1}(y|x,\ell)p_{k/t}(y,\ell)}{p_{k/k-1}(y,\ell)}dy\right)}{\int p_{k-1/k-1}(x, \ell)\left(\alpha_{s,k/t}(x, \ell) + \beta_{s,k/t}(x, \ell)\int \frac{f_{k/k-1}(y|x,\ell)p_{k/t}(y,\ell)}{p_{k/k-1}(y,\ell)}dy\right)dx} \tag{10}$$

$\alpha_{s,k/t}(x, \ell)$ and $\beta_{s,k/t}(x, \ell)$ are defined as

$$\alpha_{s,k/t}(x, \ell) \triangleq \frac{\left(1 - r^\ell_{k/t}\right)\left(1 - p_{s,k/k-1}(x, \ell)\right)}{\left(1 - r^\ell_{k/k-1}\right)} \tag{11}$$

$$\beta_{s,k/t}(x, \ell) \triangleq \frac{r^\ell_{k/t}p_{s,k/k-1}(x, \ell)}{r^\ell_{k/k-1}} \tag{12}$$

Proof of Proposition 1: From the assumptions, $\pi_{k/t}(\mathbf{Y})$, $\pi_{k/k-1}(\mathbf{Y})$ and $\pi_{k-1/k-1}(\mathbf{X})$ are LMB RFS and can be represented as

$$\pi_{k/t}(\mathbf{Y}) = \Delta(\mathbf{Y})\left(\prod_{\ell \in \mathbb{L} \uplus \mathbb{B}}\left(1 - r^\ell_{k/t}\right)\prod_{\ell \in L(\mathbf{Y})}\frac{1_{\mathbb{L} \uplus \mathbb{B}}(\ell)r^\ell_{k/t}}{\left(1 - r^\ell_{k/t}\right)}\right)\left[p_{k/t}(y, \ell)\right]^{\mathbf{Y}} \tag{13}$$

$$\pi_{k/k-1}(\mathbf{Y}) = \Delta(\mathbf{Y}) \left(\prod_{\ell \in \mathbb{L} \uplus \mathbb{B}} \left(1 - r^\ell_{k/k-1} \right) \prod_{\ell \in L(\mathbf{Y})} \frac{1_{\mathbb{L} \uplus \mathbb{B}}(\ell) r^\ell_{k/k-1}}{\left(1 - r^\ell_{k/k-1} \right)} \right) \left[p_{k/k-1}(y, \ell) \right]^{\mathbf{Y}} \tag{14}$$

$$\pi_{k-1/k-1}(\mathbf{X}) = \Delta(\mathbf{X}) \left(\prod_{\ell \in \mathbb{L}} \left(1 - r^\ell_{k-1/k-1} \right) \prod_{\ell \in L(\mathbf{X})} \frac{1_{\mathbb{L}}(\ell) r^\ell_{k-1/k-1}}{\left(1 - r^\ell_{k-1/k-1} \right)} \right) \left[p_{k-1/k-1}(x, \ell) \right]^{\mathbf{X}} \tag{15}$$

The label space of $\pi_{k/k-1}(\mathbf{Y})$ is $\mathbb{L} \uplus \mathbb{B}$ where $\mathbb{L} = \mathbb{L}_{1:k-1}$ and $\mathbb{B} = \mathbb{L}_k$. $\pi_{k/t}(\mathbf{Y})$ is initialized with $\pi_{t/t}(\mathbf{Y})$ which is the multi-target posterior density at moment t and is approximated by a LMB RFS so the label space of $\pi_{k/t}(\mathbf{Y})$ is $\mathbb{L} \uplus \mathbb{B}$ when $k = t$. $\pi_{k/t}(\mathbf{Y})$ keeps the same label space with $\pi_{k/k-1}(\mathbf{Y})$ at the other moments and the reason will be explained in (20). The density of the newborn targets is a LMB RFS and can be represented as $\pi_{B,k/k-1}(\mathbf{Y}) = \left\{ \left(r^\ell_{B,k/k-1}, p^\ell_{B,k/k-1} \right) \right\}_{\ell \in \mathbb{B}}$. The multi-target Markov transition density $f_{k/k-1}(\mathbf{Y}|\mathbf{X})$ can be represented by (7).

Let $\mathbf{Y} = \mathbf{Y}^+ \uplus \mathbf{Y}^-$. $\mathbf{Y}^+$ represents the newborn targets generated by $\pi_{B,k/k-1}(\mathbf{Y}^+)$ at moment k and $L(\mathbf{Y}^+) \subset \mathbb{B}$. $\mathbf{Y}^-$ represents the surviving targets from moment $k - 1$ to moment k and $L(\mathbf{Y}^-) \subset \mathbb{L}$. Combine (7) and (13)–(15), the backward smoothing Eq. (4) can be transformed into

$$\pi_{k-1/t}(\mathbf{X}) = \pi_{k-1/k-1}(\mathbf{X}) \int f_{B,k/k-1}(\mathbf{Y}^+) f_{s,k/k-1}(\mathbf{Y}^-|\mathbf{X}) \frac{\pi_{k/t}(\mathbf{Y})}{\pi_{k/k-1}(\mathbf{Y})} \delta\mathbf{Y}$$

$$= \pi_{k-1/k-1}(\mathbf{X}) \underbrace{\int f_{s,k/k-1} \frac{\pi^-_{k/t}(\mathbf{Y}^-)}{\pi^-_{k/k-1}(\mathbf{Y}^-)} \delta\mathbf{Y}^-}_{\text{Survived targets}} \underbrace{\int f_{B,k/k-1}(\mathbf{Y}^+) \frac{\pi^+_{k/t}(\mathbf{Y}^+)}{\pi^+_{k/k-1}(\mathbf{Y}^+)} \delta\mathbf{Y}^+}_{\text{Born targets}} \tag{16}$$

Where

$$\pi_{k/t}(\mathbf{Y}) = \pi^-_{k/t}(\mathbf{Y}^-) \pi^+_{k/t}(\mathbf{Y}^+) \tag{17}$$

$$\pi^-_{k/t}(\mathbf{Y}^-) = \Delta(\mathbf{Y}^-) \left(\prod_{\ell \in \mathbb{L}} \left(1 - r^\ell_{k/t} \right) \prod_{\ell \in L(\mathbf{Y}^-)} \frac{1_{\mathbb{L}}(\ell) r^\ell_{k/t}}{\left(1 - r^\ell_{k/t} \right)} \right) \left[p_{k/t}(y, \ell) \right]^{\mathbf{Y}^-} \tag{18}$$

$$\pi^+_{k/t}(\mathbf{Y}^+) = \Delta(\mathbf{Y}^+) \left(\prod_{\ell \in \mathbb{B}} \left(1 - r^\ell_{k/t} \right) \prod_{\ell \in L(\mathbf{Y}^+)} \frac{1_{\mathbb{B}}(\ell) r^\ell_{k/t}}{\left(1 - r^\ell_{k/t} \right)} \right) \left[p_{k/t}(y, \ell) \right]^{\mathbf{Y}^+} \tag{19}$$

The decomposition of (17) is applied the formula (3) and $\pi_{k/k-1}(\mathbf{Y})$ is also applied the similar decomposition. The derivation of (16) applies the proposition that the single set integral on the joint space can be represented as a multiple set integral on the disjoint subspaces in *Section* 3.5.3 of [1]. The formula (16) contains two parts: the first part relates to $\mathbf{X}$ which is the smoothing of survived targets, and the second part is the smoothing of the birth targets. As $f_{B,k/k-1}(\mathbf{Y}^+) = \pi_{k/k-1}(\mathbf{Y}^+)$, the second part of (16) is equal to 1 and the result can be obtained from

$$\int f_{B,k/k-1}(\mathbf{Y}^+) \frac{\pi_{k/t}^+(\mathbf{Y}^+)}{\pi_{k/k-1}^+(\mathbf{Y}^+)} \delta \mathbf{Y}^+ = \int \pi_{k/t}^+(\mathbf{Y}^+)\delta \mathbf{Y}^+ = 1 \qquad (20)$$

The formula (20) can be explained further: if a target is born at moment k, it cannot be alive at moment $k-1$, thus we don't need to consider the newborn targets for backward smoothing. Furthermore, we explain the label space of (13). Newborn targets cannot transmit to the moment before target birth, and $\pi_{k/t}(\mathbf{Y})$ cannot have the targets born after moment k, so $\pi_{k/t}(\mathbf{Y})$ and $\pi_{k/k-1}(\mathbf{Y})$ have the same label space $\mathbb{L} \uplus \mathbb{B}$ which is the union of label space at moment $k-1$ (including the label spaces prior to moment $k-1$) and label space of newborn targets at moment k.

The term $\pi_{k/t}^-(\mathbf{Y}^-)\big/\pi_{k/k-1}^-(\mathbf{Y}^-)$ in (16) can be written as

$$\frac{\pi_{k/t}^-(\mathbf{Y}^-)}{\pi_{k/k-1}^-(\mathbf{Y}^-)} = \Delta(\mathbf{Y}^-)\prod_{\ell\in\mathbb{L}}\frac{\left(1-r_{k/t}^\ell\right)}{\left(1-r_{k/k-1}^\ell\right)}\prod_{\ell\in L(\mathbf{Y}^-)}\frac{1_{\mathbb{L}}(\ell)r_{k/t}^\ell\left(1-r_{k/k-1}^\ell\right)p_{k/t}(y,\ell)}{\left(1-r_{k/t}^\ell\right)r_{k/k-1}^\ell p_{k/k-1}(y,\ell)} \qquad (21)$$

Combine (11)–(12) and (20)–(21), the formula (16) can be transformed into

$$
\begin{aligned}
\pi_{k-1/t}(\mathbf{X}) &= \pi_{k-1/k-1}(\mathbf{X})\int f_{s,k/k-1}(\mathbf{Y}^-|\mathbf{X})\frac{\pi_{k/t}^-(\mathbf{Y}^-)}{\pi_{k/k-1}^-(\mathbf{Y}^-)}\delta\mathbf{Y}^- \\
&= \pi_{k-1/k-1}(\mathbf{X})\left(\frac{1-r_{k/t}^\ell}{1-r_{k/k-1}^\ell}\right)^{\mathbb{L}-L(\mathbf{X})}\left(\alpha_{s,k/t}(x,\ell)\right)^{\mathbf{X}}\int\Delta(\mathbf{Y}^-)1_{L(\mathbf{X})}(L(\mathbf{Y}^-)) \\
&\quad\times\left(\frac{\beta_{s,k/t}(x,\ell)}{\alpha_{s,k/t}(x,\ell)}\right)^{L(\mathbf{Y}^-)}\left(\frac{f_{k/k-1}(y|x,\ell)p_{k/t}(y,\ell)}{p_{k/k-1}(y,\ell)}\right)^{\mathbf{Y}^-}\delta\mathbf{Y}^- \\
&= \pi_{k-1/k-1}(\mathbf{X})\left(\frac{1-r_{k/t}^\ell}{1-r_{k/k-1}^\ell}\right)^{\mathbb{L}-L(\mathbf{X})}\left(\alpha_{s,k/t}(x,\ell)\right)^{\mathbf{X}} \\
&\quad\times\sum_{L(\mathbf{Y}^-)\subset L(\mathbf{X})}\prod_{\ell\in L(\mathbf{Y}^-)}\int\frac{\beta_{s,k/t}(x,\ell)f_{k/k-1}(y|x,\ell)p_{k/t}(y,\ell)}{\alpha_{s,k/t}(x,\ell)p_{k/k-1}(y,\ell)}dy \\
&= \pi_{k-1/k-1}(\mathbf{X})\left(\frac{1-r_{k/t}^\ell}{1-r_{k/k-1}^\ell}\right)^{\mathbb{L}-L(\mathbf{X})}\left(\alpha_{s,k/t}(x,\ell)\right)^{\mathbf{X}} \\
&\quad\times\prod_{\ell\in L(\mathbf{X})}\left(1+\int\frac{\beta_{s,k/t}(x,\ell)f_{k/k-1}(y|x,\ell)p_{k/t}(y,\ell)}{\alpha_{s,k/t}(x,\ell)p_{k/k-1}(y,\ell)}dy\right) \\
&= \Delta(\mathbf{X})\prod_{\ell\in\mathbb{L}-L(\mathbf{X})}\frac{\left(1-r_{k-1/k-1}^\ell\right)\left(1-r_{k/t}^\ell\right)}{\left(1-r_{k/k-1}^\ell\right)}\prod_{\ell\in L(\mathbf{X})}\left(1_{\mathbb{L}}(\ell)r_{k-1/k-1}^\ell\right. \\
&\quad\times\left. p_{k-1/k-1}(x,\ell)\left(\alpha_{s,k/t}(x,\ell)+\int\beta_{s,k/t}(x,\ell)\frac{f_{k/k-1}(y|x,\ell)p_{k/t}(y,\ell)}{p_{k/k-1}(y,\ell)}dy\right)\right)
\end{aligned}
\qquad (22)
$$

The third step in (22) applies *Lemma 1* in *Section* 15.5.1 of [1]. The fourth step in (22) applies the power-functional identity in *Section* 3.7 of [1]. In the last step of (22), the formula (15) is substituted. Then let

$$
\begin{aligned}
\pi_{k-1/t}(x,\ell) &= r^{\ell}_{k-1/t} p_{k-1/t}(x,\ell) \\
&= r^{\ell}_{k-1/k-1} p_{k-1/k-1}(x,\ell) \times \\
&\quad \left(\alpha_{s,k/t}(x,\ell) + \int \frac{\beta_{s,k/t}(x,\ell) f_{k/k-1}(y|x,\ell) p_{k/t}(y,\ell)}{p_{k/k-1}(y,\ell)} dy \right)
\end{aligned}
\tag{23}
$$

From (23), we can obtain

$$
\begin{aligned}
r^{\ell}_{k-1/t} &= \int \pi_{k-1/t}(x,\ell) dx \\
&= \int \alpha_{s,k/t}(x,\ell) r^{\ell}_{k-1/k-1} p_{k-1/k-1}(x,\ell) dx + \\
&\quad \iint \frac{r^{\ell}_{k-1/k-1} p_{k-1/k-1}(x,\ell) \beta_{s,k/t}(x,\ell) f_{k/k-1}(y|x,\ell) p_{k/t}(y,\ell)}{p_{k/k-1}(y,\ell)} dy dx
\end{aligned}
\tag{24}
$$

Where the second term of the right of (24) can be transformed into

$$
\begin{aligned}
&\iint \frac{r^{\ell}_{k-1/k-1} p_{k-1/k-1}(x,\ell) \beta_{s,k/t}(x,\ell) f_{k/k-1}(y|x,\ell) p_{k/t}(y,\ell)}{p_{k/k-1}(y,\ell)} dy dx \\
&= \int \frac{\int r^{\ell}_{k-1/k-1} p_{k-1/k-1}(x,\ell) p_{s,k/k-1}(x,\ell) f_{k/k-1}(y|x,\ell) dx \, r^{\ell}_{k/t} p_{k/t}(y,\ell)}{r^{\ell}_{k/k-1}} \frac{1}{p_{k/k-1}(y,\ell)} dy \\
&= \int r^{\ell}_{k/t} p_{k/t}(y,\ell) dy \\
&= r^{\ell}_{k/t}
\end{aligned}
\tag{25}
$$

And the first term of the right of (24) can be transformed into

$$
\begin{aligned}
&\int \alpha_{s,k/t}(x,\ell) r^{\ell}_{k-1/k-1} p_{k-1/k-1}(x,\ell) dx \\
&= \frac{\left(1 - r^{\ell}_{k/t}\right) r^{\ell}_{k-1/k-1}}{\left(1 - r^{\ell}_{k/k-1}\right)} \int \left(1 - p_{s,k/k-1}(x,\ell)\right) p_{k-1/k-1}(x,\ell) dx \\
&= \frac{\left(1 - r^{\ell}_{k/t}\right)}{\left(1 - r^{\ell}_{k/k-1}\right)} \left(r^{\ell}_{k-1/k-1} - r^{\ell}_{k/k-1} \right)
\end{aligned}
\tag{26}
$$

Combine (25–26) and (23)–(24), we can obtain (9) and (10) respectively. From (9), it is known that the smoothed existence probability of the label ℓ relates to $r_{k-1/k-1}$,

$r_{k/k-1}$ and $r_{k/\ell}$. From (10), it is known that the density of the label ℓ contain two parts where one part only relates to $p_{k-1/k-1}(x, \ell)$ and $\alpha_{s,k/t}(x, \ell)$ then another part relates to the backward smoothing.

Substitute (23) into (22), the formula (22) can be represented as

$$
\begin{aligned}
\pi_{k-1/t}(\mathbf{X}) &= \Delta(\mathbf{X}) \prod_{\ell \in \mathbb{L}-L(\mathbf{X})} \left(1 - r_{k-1/t}^\ell\right) \prod_{\ell \in L(\mathbf{X})} \left(1_{\mathbb{L}}(\ell) r_{k-1/t}^\ell p_{k-1/t}(x, \ell)\right) \\
&= \left\{ \left(r_{k-1/t}^\ell, p_{k-1/t}(x, \ell)\right) \right\}_{\ell \in \mathbb{L}}
\end{aligned}
\tag{27}
$$

The formula (27) shows that $\pi_{k-1/t}(\mathbf{X})$ is a LMB distribution where $r_{k-1/t}^\ell$ and $p_{k-1/t}(x, \ell)$ are represented by (9) and (10) respectively. The backward smoothed density of each target is independent and the smoothed density $\left(r_{k-1/t}^\ell, p_{k-1/t}(x, \ell)\right)$ of the target (x, ℓ) is consistent with the smoothed Bernoulli density in [6] when the newborn target is not considered.

The LMB smoother is simple and effective for the two reasons. Firstly, the newborn targets are uncorrelated with the backward smoothing since the newborn target cannot be alive prior to the birth time. Secondly, the existence probabilities and the probability densities of different targets are uncorrelated for the LMB smoother.

4 Conclusions

The paper derives a forward-backward LMB smoother. Applying the conclusion that the newborn targets do not influence the backward smoothing, we prove that the backward smoothing is closed under the LMB prior and the backward smoothed density of each target is consistent with the smoothed Bernoulli density in [6]. As the LMB smoother here has a simple and effective form than GLMB smoother, the subsequent work is implementing and evaluating it by some experiments.

References

1. Mahler, R.P.: Advances in Statistical Multisource Multitarget Information Fusion. Artech House, Norwood (2014)
2. Reuter, S., et al.: The labeled multi-Bernoulli filter. IEEE Trans. Sig. Process. **62**(12), 3246–3260 (2014)
3. Mahler, R.P.S., Vo, B.T.: Forward-backward probability hypothesis density smoothing. IEEE Trans. Aerosp. Electron. Syst. **48**(1), 707–728 (2012)
4. Mahler, R.P.S., Vo, B.N., Vo, B.T.: The forward-backward probability hypothesis density smoother. In: Information Fusion (2011)
5. Vo, B.N., Vo, B.T., Mahler, R.P.S.: Closed-form solutions to forward-backward smoothing. IEEE Trans. Sig. Process. **60**(1), 2–17 (2011)
6. Vo, B.T., et al.: Bernoulli forward-backward smoothing for joint target detection and tracking. IEEE Trans. Sig. Process. **59**(9), 4473–4477 (2011)

7. Dong, L., Hou, C., Yi, D.: Multi-Bernoulli smoother for multi-target tracking. Aerosp. Sci. Technol. **48**, 234–245 (2016)
8. Beard, M., Vo, B.T., Vo, B.N.: Generalised labelled multi-Bernoulli forward-backward smoothing. In: International Conference on Information Fusion (2016)
9. Tiancheng, L., et al.: Joint smoothing and tracking based on continuous-time target trajectory function fitting. IEEE Trans. Autom. Sci. Eng. 1–8 (2018)

Kullback-Leibler Averaging
for Multitarget Density Fusion

Kai Da[1](✉), Tiancheng Li[2], Yongfeng Zhu[1], Hongqi Fan[1], and Qiang Fu[1]

[1] National Key Laboratory of Science and Technology on ATR,
College of Electronic Science, National University of Defense Technology,
Changsha 410073, China
dktm131@163.com, zoyofo@163.com, fanhongqi@nudt.edu.cn,
fuqiang1962@vip.sina.com
[2] Key Laboratory of Information Fusion Technology (Ministry of Education),
School of Automation, Northwestern Polytechnical University,
Xi'an 710072, China
t.c.li@mail.nwpu.edu.cn, t.c.li@usal.es

Abstract. This paper addresses the linear and log-linear fusion approaches to multitarget density fusion which yield arithmetic average (AA) and geometric average (GA), respectively. We reaffirm Abbas's finding in 2009 that both AA and GA can be related to the minimization of the Kullback-Leibler divergence (KLD) between the fusing densities and the fused result, which differ from each other in the reference used to measure the KLD: the AA uses the fusing densities while the GA uses the fused density. We derive the explicit AA expressions for fusing some known multitarget densities and discuss the implementation issues. The results serve as the theoretical basis for designing distributed random finite set filters for distributed multitarget tracking.

Keywords: Average consensus · Arithmetic average · Linear fusion ·
Random finite set · Sensor network · Target tracking

1 Introduction

Multisensor multitarget tracking in the presence of false, missing and noisy data [1,2] has led to substantial research interest in both military and commercial realms, for which a significant scientific problem is multisensor data fusion. Particular interest has been paid to calculating the "average" over the information provided by local sensors/agents based on either linear or log-linear fusion, which yields the arithmetic average (AA) and geometric average (GA) [3], respectively. The research field of "average consensus" [4,5] has burgeoned with the vitalization of networked systems during the last years.

As earlier noticed by Abbas in 2009 [6], the linear fusion minimizes the sum of the Kullback-Leibler divergences (KLDs) from the fusing densities to the fused density while the log-linear fusion minimizes that from the fused density to the

© Springer Nature Switzerland AG 2020
F. Herrera et al. (Eds.): DCAI 2019, AISC 1003, pp. 253–261, 2020.
https://doi.org/10.1007/978-3-030-23887-2_29

fusing densities. While the GA/log-linear fusion has been well understood and used in the Bayesian inference community, much less attention has been paid to the AA fusion. However, it has been recently demonstrated that the AA fusion employed for designing average consensus algorithms can yield better performance as compared with the GA fusion in fusing multitarget densities (MTDs) which are contracted with false alarms and missed detections, and significantly different from the standard single target density [3, 7–11].

This paper focuses on the theoretical study of the AA fusion and is organized as follows: Preliminaries are given in Sect. 2. We prove the KLD-minimization property of the AA fusion based on set integral and derive the MTD-AA results for several well known MTDs in Sects. 3 and 4, respectively. Our comments on implementing the MTD-AA fusion are provided in Sect. 5.

2 Preliminaries

2.1 Set Integral and Probability Hypothesis Density

For the MTD function $f(X)$ of the multitarget set $X = \{x_1, ..., x_n\}$, the set integral [2] in a region S is given

$$\int_S f(X)\delta X = f(\emptyset) + \sum_{n=1}^{\infty} \frac{1}{n!} \int_{S^n} f(\{x_1, ..., x_n\})dx_1...dx_n. \tag{1}$$

The cardinality distribution $p(n)$ of the random set X can be computed as

$$p(n) = \int_{|X|=n} f(X)\delta X = \frac{1}{n!} \int f(\{x_1, ..., x_n\})dx_1...dx_n. \tag{2}$$

The probability generating functional (p.g.fl.) is equivalent form of the multitarget process. The p.g.fl. of the random set X is defined as $G[h] = E[h^X] = \int h^X f(X)\delta X$, where $h^X = \begin{cases} 1, X = \emptyset \\ \prod_{x \in X} h(x), X \neq \emptyset \end{cases}$ and $0 \leq h(x) \leq 1$ is an unitless test function. The probability hypothesis density (PHD) $D(x)$ of multitarget density is computed by the p.g.fl and by set integral as follows, respectively

$$D(x) = \left. \frac{\delta G}{\delta x}[h] \right|_{h=1}, \tag{3}$$

$$D(x) = \int f(\{x\} \cup W)\delta W. \tag{4}$$

2.2 Kullback-Leibler Divergence

The Kullback-Leibler divergence (KLD) is the most used metric to measure the divergence between distributions. The KLD from distribution $f(X)$ to distribution $g(X)$ is given as

$$D_{KL}(f(X)\|g(X)) = \int f(X)\log(f(X)/g(X))\,\delta X, \tag{5}$$

which is a measure of the information gained when $f(X)$ is used to replace $g(X)$, or to say, a measure of the information lost when $g(X)$ is used to approximate $f(X)$ [12, p. 51]. In this paper the integral is a set integral as the concerning distributions are MTDs [13]. The KLD $D_{\mathrm{KL}}(f(X)\|g(X))$ is no-negative and is zero if and only if $f(X) = g(X)$. Further on, the KLD is directional, i.e., $D_{\mathrm{KL}}(f(X)\|g(X)) \neq D_{\mathrm{KL}}(g(X)\|f(X))$. We refer to $f(X)$ in (5) as *the reference distribution* of the KLD. Either $f(X)$ or $g(X)$ can be used as the reference for evaluating the divergence between them which leads to different results.

2.3 GA and AA Formula

Given the MTD $f_i(X)$ from the ith sensor, the AA is given by

$$f_{\mathrm{AA}}(X) = \sum_i w_i f_i(X), \tag{6}$$

where $w_i > 0$ is the fusing weight and the average requires $\sum_i w_i = 1$.

In contrast, the density GA is given by

$$f_{\mathrm{GA}}(X) = \left(\int \prod_i (f_i(Y))^{w_i} \delta Y \right)^{-1} \prod_i (f_i(X))^{w_i}. \tag{7}$$

3 Weighted Kullback-Leibler Average for Consensus

The standard average consensus algorithm is defined over variables [4,5]. We extend it to accommodate MTDs. To this purpose, both the definition of the average and the metric used for evaluating the quality of consensus need to be revised. While the quality of consensus in the former is usually measured by the total mean-square deviation of the individual variables from their average [5] and similar, we use the weighted sum of the KLDs between the fusing MTDs and the fused MTD as the cost function to be minimized.

Given local densities $f_i(X), i = 1, ..., N$, the global average (over all sensors) and regional average for sensor i are given as follows, respectively,

$$\bar{f}(X) = \mathcal{M}\left(\{w_i, f_i(X)\}_{i=1,2,...,N} \right), \bar{f}_{i,l+1}(X) = \mathcal{M}\left(\{w_{i \leftarrow j}, \bar{f}_{j,l}(X)\}_{j \in \mathcal{N}^i} \right), \tag{8}$$

where $\mathcal{M}(\cdot)$ is an averaging function, N is the total number of sensors, $\mathcal{N}^i$ is the set of neighbors of sensor i including itself, l denotes the consensus iteration and $w_{i \leftarrow j}$ is the weight assigned to $\bar{f}_j(X)$ for fusion at sensor i , satisfying $\sum w_j = 1$ and $\bar{f}_{j,0}(X) = f_j(X)$. The asymptotic convergence of the average consensus algorithm [4,5] states

$$\lim_{l \to \infty} \bar{f}_{i,l}(X) = \bar{f}(X). \tag{9}$$

In what follows, we investigate the definition of the averaging function $\mathcal{M}(\cdot)$ corresponding to the GA and AA, respectively.

Theorem 1. *Based on the definitions* (6), (7) *and* (5), *we have*

$$f_{AA}(X) = \arg\min_{g(X)} \sum_i w_i D_{KL}(f_i(X)\|g(X)), \tag{10}$$

$$f_{GA}(X) = \arg\min_{g(X)} \sum_i w_i D_{KL}(g(X)\|f_i(X)). \tag{11}$$

Proof. (11) has been exactly proved in [13] and following the same line of the proof, we can easily derive (10). That is, the cost to be minimized in (10) is

$$J(g(X)) \triangleq \sum_i w_i D_{KL}(f_i(X)\|g(X)) = \sum_i w_i \int f_i(X) \log \frac{f_i(X)}{g(X)} \delta X$$

$$= \int \sum_i w_i f_i(X) \log f_i(X) - f_{AA}(X) \log f_{AA}(X) \delta X$$

$$+ \int f_{AA}(X) \log f_{AA}(X) - f_{AA}(X) \log g(X) \delta X$$

$$= \sum_i w_i D_{KL}(f_i(X)\|f_{AA}(X)) + D_{KL}(f_{AA}(X)\|g(X)), \tag{12}$$

Since the KLD is always non-negative and is zero if and only if its two arguments coincide almost everywhere, the above cost is trivially minimized by taking $g(X) = f_{AA}(X)$. So, (10) was also proved.

Remark 1. *Both AA* (10) *and GA* (11) *fusion approaches minimize the weighted KLDs, differing from each other only in the reference used for calculating the KLD. Both can be referred to as the Kullback-Leibler average (KLA) and interpreted as that, they are the consensus that all fusing agents can achieve at the price of "making the least change". While the GA fusion uses the fused MTD as the reference MTD for evaluating the information change of each local MTD, the AA fusion uses the local original MTDs as the reference.*

Switching the KLD reference from the fused distribution to the fusing distributions facilitates easier calculation [14]. Abbas [6] was the first noticing the bidirectional results as in (10) and in (11). In parallel to our work, we have noticed that the bidirectional KLD results have also been noticed and independently extended to multitarget density fusion by [11].

Remark 2. *The KLD-minimization property of either the AA or GA fusion, however, does not ensure benefiting the filter in accuracy (except the fusing items are i.i.d. and all unbiased for which the AA is statistically more accurate [8]) but instead, either of them may make things worse in practice [3]; see for example the experimental evidence on the AA fusion [15] and on the GA fusion [9, 11].*

4   AA Fusion for Multitarget Densities

While the exact closed forms of the GA fusions for several popular RFS densities have been derived in [16], little attention has been paid to AA fusion although it has superiority in dealing with missed detection and in analytically fusing mixtures [3]. To fill this gap, in what follows we investigate the AA fusion for various multitarget densities.

4.1 AA Fusion of Bernoulli Posterior

The Bernoulli filter [17] provides an optimal Bayesian method in the single target tracking problem.

Theorem 2. *Assume that the posterior of the ith sensor is Bernoulli process, then,*

$$f_i(X|Z_i^k) = \begin{cases} 1 - p_i, & X = \emptyset \\ p_i s_i(x), & X = \{x\}, \end{cases} \tag{13}$$

where p_i is the target existence probability and $s_i(x)$ is the spatial density for single target. The fused target existence probability and the track density are

$$p_{AA} = \sum_i w_i p_i, \quad s_{AA}(x) = \left(\sum_i w_i p_i s_i(x) \right) / p_{AA} \tag{14}$$

Proof. For the cases with $X = \emptyset$ and $X = \{x\}$, we have

$$f_{AA}(\emptyset|Z_1^k, Z_2^k, ...) = \sum_i w_i f_i(\emptyset|Z_i^k) = \sum_i w_i(1 - p_i), \tag{15}$$

$$f_{AA}(\{x\}|Z_1^k, Z_2^k, ...) = \sum_i w_i f_i(\{x\}|Z_i^k) = \sum_i w_i p_i s_i(x). \tag{16}$$

Then the fused target existence probability and the track density are given by, respectively

$$p_{AA} = 1 - f_{AA}(\emptyset|Z_1^k, Z_2^k, ...), \quad s_{AA}(x) = \left(f_{AA}(\{x\}|Z_1^k, Z_2^k, ...) \right) / p_{AA}. \tag{17}$$

4.2 AA Fusion of Poisson Multi-target Intensities

The Poisson process is considered in PHD filter which propagates the first-order moment of the multitarget density (namely PHD).

Theorem 3. *Assume that the multitarget posterior from the ith sensor is Poisson process, then, the MTD and PHD of sensor i are*

$$f_i(X|Z_i^k) = e^{-\mu_i} \prod_{x \in X} \mu_i s_i(x), \quad D_i(x) = \mu_i s_i(x). \tag{18}$$

where μ_i is the mean of the Poisson distribution or the average number of the detected targets. The fused PHD using AA fusion is given by

$$D_{AA}(x) = \sum_i w_i D_i(x). \tag{19}$$

Proof. The p.g.fl. of the multitarget random set X is

$$G_{AA}[h] = \int h^X \sum_i w_i f_i(X|Z_i^k) \delta X$$

$$= \sum_{n=0}^{\infty} \frac{1}{n!} \int \prod_j h(x_j) \left[\sum_i w_i e^{-\mu_i} \prod_j \mu_i s_i(x_j) \right] dx_1 ... dx_n$$

$$= \sum_i w_i e^{-\mu_i} \sum_{n=0}^{\infty} \frac{1}{n!} \left(\int \mu_i h(x) s_i(x) dx \right)^n = \sum_i w_i e^{-\mu_i + \int \mu_i h(x) s_i(x) dx}.$$

$$\tag{20}$$

Using (3), the fused PHD is computed by

$$D_{\mathrm{AA}}(x) = \frac{\delta}{\delta x} G_{\mathrm{AA}}[h]\Big|_{h=1} = \left(\sum_i w_i \mu_i h(x) s_i(x) e^{-\mu_i + \int \mu_i h(x) s_i(x) dx} \right)\Big|_{h=1}$$
$$= \sum_i w_i \mu_i s_i(x) = \sum_i w_i D_i(x) \qquad (21)$$

4.3   AA Fusion of i.i.d. Cluster Intensities

The Cardinalized PHD (CPHD) filter [2,18] which simultaneously propagates the PHD and the cardinality distribution uses the i.i.d. cluster processes to approximate the multitarget processes.

Theorem 4. *Assume that the multitarget posterior of the ith sensor is i.i.d. process, then, the MTD and PHD of sensor i are*

$$f_i(X|Z_i^k) = n! p_i(n) \prod_{x \in X} s_i(x), \quad D_i(x) = \sum_{n=0}^{\infty} n p_i(n) s_i(x). \qquad (22)$$

where $p_i(n)$ is the cardinality distribution of sensor i. The fused PHD and cardinality distribution are given by

$$D_{AA}(x) = \sum_i w_i D_i(x), \quad p_{AA}(n) = \sum_i w_i p_i(n). \qquad (23)$$

Proof. The fused PHD is calculated through (4), i.e.

$$D_{\mathrm{AA}}(x) = \int \sum_i w_i f_i(\{x\} \cup X) \delta X \qquad (24)$$
$$= \sum_{n=0}^{\infty} \frac{1}{n!} \int \left\{ \sum_i w_i (n+1)! p_i(n+1) s_i(x) \prod_{x_j \in X} s_i(x_j) \right\} dx_1 ... dx_n$$
$$= \sum_{n=0}^{\infty} \left\{ \sum_i w_i \cdot (n+1) \cdot p_i(n+1) \cdot s_i(x) \right\} = \sum_i w_i D_i(x).$$

The cardinality distribution is computed with (cf. (2))

$$p_{\mathrm{AA}}(n) = \int_{|X|=n} \sum_i w_i f_i(X) \delta X = \frac{1}{n!} \int \sum_i w_i n! p_i(n) \prod_{x_j \in X} s_i(x_j) dx_1 ... dx_n$$
$$= \sum_i w_i p_i(n) \qquad (25)$$

4.4   AA Fusion of Multi-Bernoulli Posterior

The multitarget multi-Bernoulli (MB) densities [2,19] from sensor i is

$$f_{i,\mathrm{MB}}(X) = \left(\prod_{n=1}^{M_i} (1 - p_{i_n}) \right) \sum_{1 \le i_1 \ne \cdots \ne i_n \le M_i} \frac{p_{i_1} s_{i_1}(x)}{1 - p_{i_1}} \cdots \frac{p_{i_n} s_{i_n}(x)}{1 - p_{i_n}}, \qquad (26)$$

where M_i is the number of Bernoulli component, p_{i_n} and $s_{i_n}(x)$ are the existence probability and the spatial density of i_nth component, respectively.

The AA fusion of MB process is

$$f_{\mathrm{AA,MB}}(X) = \sum_i w_i f_{i,\mathrm{MB}}(X), \tag{27}$$

which, however, is no longer the MB process. That is, the AA fusion does not comply with the desired MB-closure recursion. To address this non-closure problem, three solutions may be taken. The first is to fuse after association, i.e., the Bernoulli components in one sensors are associated with that in other sensors, enable the closure fusion to be performed with respect to Bernoulli process fusion as addressed in Sect. 4.1. The second solution is to find a reasonable MB process approximation to $f_{\mathrm{AA,MB}}(X)$ (with minor KLD). The last choice is just to disregard the non-closure problem, simply mix/combine the re-weighted Gaussian mixture (GM) or particles (which are used to represent the MBs) for AA fusion and treat the fused result as a MB process.

5   Implementation Issues

We have extensively addressed the MTD-AA-fusion-based distributed PHD filters with the use of either the GM [7] or the particle filter [9], based on either the standard consensus paradigm or the flooding paradigm [20]. The implementations of the AA fusion for (labeled) multi-Bernoulli/Bernoulli processes are still missing so far and form our future work. What follows outlines our suggestions for effective implementation of the MTD-AA fusion.

1. Since the AA fusion preserves all components in the GM or all particles in the particle filter which leads to increased uncertainty of the fused distribution [3], it is important and beneficial to perform "partial consensus" [7,9] rather than complete consensus. That is sharing and fusing only a major part of the MTD rather than the entire. This does not only reduce the communication and computation cost but also benefit the accuracy much.
2. Operations such as mixture reduction (typically including merging and pruning) and particle resampling are crucial to the AA fusion, although from the Bayesian viewpoint they make little change to the posterior distribution. Just like both mixture reduction and resampling operations are important to the generic GM and particle filters, they are even more crucial in the AA-fusion-based distributed filters.
3. AA fusion is basically a data-driven method (e.g., most of clustering methods [20] can be viewed as (weighted) AA fusion), which is particularly appealing to the large-scale sensor networks for which the GA fusion, however, is vulnerable to missed detection and can significantly degrade [3,9,11].
4. For the labeled RFS densities, the labels should be matched across sensors before the densities fused; see the proposal for GA fusion in [21].
5. It might be interesting to investigate a hybrid of AA and GA fusion approaches, e.g., employ AA fusion for cardinality fusion while GA fusion for the distribution fusion [3], GA fusion for well distant targets while AA fusion for the intractable closely-distributed target fusion.

6 Conclusion

This paper has derived the explicit AA results for fusing Bernoulli, Poisson, i.i.d. cluster and Multi-Bernoulli multitarget densities, respectively. The results serve as the theoretical basis for designing distributed random finite set filtering approaches to distributed multitarget tracking. We found that, both the AA and the GA can be interpreted as that, they are the consensus that all fusing agents achieve at the price of "making the least change" in the sense of KLD. This least-divergence property, however, does not ensure filter benefit in accuracy but instead, both averaging methods may make things worse. For the AA fusion, appropriate mixture reduction and resampling operations are highly crucial to avoid the fusion-is-worse case and are, therefore, the key to the practical implementation of AA-fusion-based distributed filters.

Acknowledgments. This work is supported by the Joint Fund of Equipment development and Aerospace Science and Technology under Grant no 6141B0624050101.

References

1. Vo, B., Mallick, M., Barshalom, Y., Coraluppi, S., Osborne, R., Mahler, R., Vo, B.: Multitarget tracking. In: Webster, J.G. (ed.) Wiley Encyclopedia of Electrical and Electronics Engineering (2015)
2. Mahler, R.P.S.: Advances in Statistical Multisource-Multitarget Information Fusion. Artech House, Norwood (2014)
3. Li, T., Fan, H., Garcia, J., Corchado, J.M.: Second-order statistics analysis and comparison between arithmetic and geometric average fusion: application to multi-sensor target tracking. Inf. Fusion **51**, 233–243 (2019)
4. Olfati-Saber, R., Fax, J.A., Murray, R.M.: Consensus and cooperation in networked multi-agent systems. Proc. IEEE **95**(1), 215–233 (2007)
5. Xiao, L., Boyd, S., Kim, S.-J.: Distributed average consensus with least-mean-square deviation. J. Parallel Distrib. Comput. **67**(1), 33–46 (2007)
6. Abbas, A.E.: A Kullback-Leibler view of linear and log-linear pools. Decis. Anal. **6**(1), 25–37 (2009)
7. Li, T., Corchado, J.M., Sun, S.: Partial consensus and conservative fusion of Gaussian mixtures for distributed PHD fusion. IEEE Trans. Aerosp. Electron. Syst. (2018, in press). https://doi.org/10.1109/TAES.2018.2882960
8. Li, T., Hlawatsch, F., Djurić, P.M.: Cardinality-consensus-based PHD filtering for distributed multitarget tracking. IEEE Sig. Process. Lett. **26**(1), 49–53 (2019)
9. Li, T., Hlawatsch, F.: A distributed SMC-PHD filter based on arithmetic average PHD fusion. arxiv: 1712.06128v2 (2018)
10. Gostar, A.K., Hoseinnezhad, R., Bab-Hadiashar, A.: Cauchy-Schwarz divergence-based distributed fusion with Poisson random finite sets. In: Proceedings of ICCAIS, Chiang Mai, Thailand, pp. 112–116 (2017)
11. Gao, L., Battistelli, G., Chisci, L.: Multiobject fusion with minimum information loss. arXiv:1903.04239, March 2019
12. Burnham, K.P., Anderson, D.R.: Model Selection and Multi-Model Inference: A Practical Information-Theoretic Approach, 2nd edn. Springer, New York (2002)

13. Battistelli, G., Chisci, L., Fantacci, C., Farina, A., Graziano, A.: Consensus CPHD filter for distributed multitarget tracking. IEEE J. Sel. Top. Sig. Proces. **7**(3), 508–520 (2013)

14. Wang, Y., Li, X.R.: A fast and fault-tolerant convex combination fusion algorithm under unknown cross-correlation. In: Proceedings of Fusion, Seattle, WA, USA, pp. 6–9, July 2009

15. Nagappa, S., Clark, D.E., Mahler, R.: Incorporating track uncertainty into the OSPA metric. In: Proceedings of FUSION 2011, Chicago, IL, USA, July 2011

16. Clark, D., Julier, S., Mahler, R., Ristic, B.: Robust multi-object sensor fusion with unknown correlations. In: Proceedings of Sensor Signal Processing for Defence (SSPD 2010), London, September 2010

17. Ristic, B., Vo, B.T., Vo, B.N., Farina, A.: A tutorial on Bernoulli filters: theory, implementation and applications. IEEE Trans. Sig. Process. **61**(13), 3406–3430 (2013)

18. Vo, B.T., Vo, B.-N., Cantoni, A.: Analytic implementations of the cardinalized probability hypothesis density filter. IEEE Trans. Sig. Process. **55**(7), 3553–3567 (2007)

19. Williams, J.L.: Marginal multi-Bernoulli filters: RFS derivation of MHT, JIPDA, and association-based MeMBer. IEEE Trans. AES **51**(3), 1664–1687 (2015)

20. Li, T., Corchado, J.M., Chen, H.: distributed flooding-then-clustering: a lazy networking approach for distributed multiple target tracking. In: Proceeding of FUSION 2018, Cambridge, UK, 10–13 July, pp. 2415–2422 (2018)

21. Li, S., Battistelli, G., Chisci, L., Yi, W., Wang, B., Kong, L.: Computationally efficient multi-agent multi-object tracking with labeled random finite sets. IEEE Trans. Sig. Process. **67**(1), 260–275 (2019)

Author Index

© Springer Nature Switzerland AG 2020
F. Herrera et al. (Eds.): DCAI 2019, AISC 1003, pp. 263–264, 2020.
https://doi.org/10.1007/978-3-030-23887-2